BARRON'S

Regents Exams and Answers

Life Science: Biology

Published by Kaplan North America, LLC, d/b/a Barron's Educational Series
1515 West Cypress Creek Road
Fort Lauderdale, Florida 33309
www.barronseduc.com

ISBN: 978-1-5062-9614-2

10 9 8 7 6 5 4 3 2 1

Contents

1. **How to Use This Book** 1

2. **Understanding the Learning Standards and the 2025 Regents Biology Exam** 5

3. **Investigations—Learning Science by Doing Science** 15

4. **The 2025 Regents Biology Exam—Format, Topics, and What to Expect** 19

5. **Test-Taking Tips for the Regents Biology Exam** 23

6. **Regents-Style Practice Clusters** 27

What's This Chapter About? 27

Topic 1: Structure and Function 29

Phenomenon: Enzymatic Activity in Digestion and Feedback Regulation in the Human Body 29

Focus: LS1.A | SEP: Constructing Explanations | CCC: Structure and Function 29

Topic 2: Matter and Energy in Organisms and Ecosystems 35

Phenomenon: Energy Flow in a Seagrass Ecosystem and the Effects of Algal Blooms 35

Focus: LS1.C | SEP: Modeling | CCC: Energy and Matter 35

Topic 3: Interdependent Relationships in Ecosystems 42

Phenomenon: Gray Wolf Reintroduction and Ecosystem Stability in Yellowstone 42

Focus: LS2.A | SEP: Using Math | CCC: Stability and Change 42

Topic 4: Inheritance and Variation of Traits 48
Phenomenon: Genetic Variation in Sickle Cell Trait and Malaria Resistance 48
Focus: LS3.A/B | SEP: Data Analysis | CCC: Cause and Effect 48

Topic 5: Natural Selection and Evolution 54
Phenomenon: Antibiotic Resistance in Bacteria 54
Focus: LS4.B/C | SEP: Constructing Explanations | CCC: Cause and Effect 54

Topic 6: Earth's Systems 60
Phenomenon: Increasing Atmospheric Carbon Dioxide and Its Impact on Climate and Ecosystems 60
Focus: ESS2.D & LS2.C | SEP: Interpreting Data | CCC: Stability and Change 60

Topic 7: Engineering, Technology, and Applications of Science 66
Phenomenon: Designing Solutions to Reduce Atmospheric CO_2 and Limit Climate Change 66
Focus: ETS1.B | SEP: Designing Solutions | CCC: Stability and Change 66

7. Regents Practice Exams 73

Regents Practice Exam June 2025 73
Answer Explanations June 2025 116
Regents Practice Exam August 2025 128
Answer Explanations August 2025 168

Chapter 1
How to Use This Book

Welcome to your complete guide for preparing for the **2025 Regents Examination in Life Science: Biology**. This isn't just a test prep book—it's your all-in-one toolkit for mastering New York State's updated science standards and excelling on the new cluster-based Regents exam.

With real-world science scenarios, Regents-style questions, and detailed explanations throughout, this book will help you learn how to **think, analyze, and explain like a scientist**—not just memorize facts.

What's Inside This Book

This book is organized into **three main parts**, designed to help you build confidence step by step—from understanding the standards to applying your knowledge on full-length exams.

Chapters 2–5: Foundations of the Regents Biology Exam

These chapters introduce everything you need to know about the 2025 Regents Biology Exam:

- **Chapter 2: Understanding the Learning Standards and the 2025 Regents Biology Exam**

  Learn what the New York State P–12 Science Learning Standards are and how they form the basis for everything on the Regents. You'll explore the three dimensions of science learning:

  - **Science and Engineering Practices (SEPs)**—what scientists do
  - **Disciplinary Core Ideas (DCIs)**—what scientists know
  - **Crosscutting Concepts (CCCs)**—how science connects across topics

- **Chapter 3: Investigations—Learning Science by Doing Science**

  Discover how Investigations—hands-on, performance-based classroom tasks—help you develop scientific thinking skills that are also tested on the Regents exam. These tasks mirror real science, support instruction throughout the year, and strengthen your ability to interpret data and explain phenomena.

- **Chapter 4: The 2025 Regents Biology Exam—Format, Topics, and What to Expect**

  Get familiar with how the test is structured: clusters, question types, and topic breakdown. You'll learn how many questions to expect, which topics are emphasized, and how to read models, graphs, and real-world data.

- **Chapter 5: Test-Taking Tips for the Regents Biology Exam**

  Build a smart strategy for exam day. Learn how to tackle clusters, manage your time, use the Claim–Evidence–Reasoning (CER) method in written responses, and avoid common mistakes.

Chapter 6: Regents-Style Practice Clusters

This is the heart of the book—**seven full practice clusters**, one for each major topic on the test. Each cluster is built to match the format of the 2025 Regents Exam and includes:

- A real-world scientific phenomenon
- One or more data-rich stimuli (graphs, diagrams, tables, or passages)
- 9–11 multiple-choice and constructed-response questions
- SEP, DCI, and CCC labeled at the top of each cluster
- Detailed answer explanations for every question

Topics include:

- Structure and Function
- Matter and Energy in Organisms and Ecosystems
- Interdependent Relationships in Ecosystems
- Inheritance and Variation of Traits
- Natural Selection and Evolution
- Earth's Systems
- Engineering and Applications of Science

Chapter 7: Regents Practice Exams

Finish your preparation with **two actual Regents practice exams**, fully aligned with the 2025 format.

Each includes:

- Stimulus-based question clusters
- A realistic mix of question types
- Detailed answer explanations

Use these exams to:

- Benchmark your readiness
- Practice under timed conditions
- Target your final areas for review

How to Work Through This Book

Follow this step-by-step approach to get the most out of your study time:

Step 1: Learn the Foundations
Read Chapters 2–5 to understand how the Regents exam works, how to use science practices, and what's expected of you.

Step 2: Practice with Clusters
Choose a topic you're learning in class and work through the related cluster. Read the phenomenon carefully, study the stimuli, and answer each question thoughtfully.

Step 3: Write and Explain
Practice writing CER-style responses. Use scientific vocabulary. Refer to graphs, models, or data when explaining your answers.

Step 4: Review Your Mistakes
Don't just check for right or wrong answers—read the explanations to understand *why* an answer is correct and learn from any mistakes.

Step 5: Take the Full Practice Exams
When you're ready, simulate test-day conditions and take the practice exams in Chapter 7. Time yourself, then review your performance.

Tips for Success

- **Think like a scientist:** Ask questions, find patterns, and explain how and why things happen.
- **Use models, graphs, and figures:** They often hold clues to answering questions correctly.
- **Practice CER writing:** Make a clear claim, back it up with data, and explain your reasoning.
- **Review SEPs, DCIs, and CCCs:** Understand what skills and concepts each question is targeting.

Final Word

This book is designed to help you succeed—not by memorizing long lists of facts, but by helping you *understand* biology and apply it to the world around you. By the end of this book, you'll be able to read like a scientist, think like a problem-solver, and write like an expert—all skills that will help you not only on the exam but also in life.

Chapter 2
Understanding the Learning Standards and the 2025 Regents Biology Exam

Welcome to your first step in preparing for the Regents Examination in Life Science: Biology. This chapter will help you understand *what* you're expected to know about biology, *why* you're learning it, and *how* it will show up on the exam. We'll walk you through the New York State P–12 Science Learning Standards (NYSP-12SLS)—the foundation for your entire biology course—and show you how the learning standards connect to the test you'll take.

Why Are There New Learning Standards?

The New York State P–12 Science Learning Standards were adopted to match national expectations and to help you develop deeper, more useful science skills. The new standards are based on the Next Generation Science Standards (NGSS), which aim to get students thinking, doing, and communicating like scientists—not just memorizing facts.

What Are the Learning Standards?

The NYSP-12SLS are built on a three-dimensional learning model, which means every topic you study in biology will include:

1. Science and Engineering Practices (SEPs)
2. Disciplinary Core Ideas (DCIs)
3. Crosscutting Concepts (CCCs)

Let's break down these topics, so you know what each means for you as a student.

Science and Engineering Practices (SEPs)

You're not just here to memorize facts—you're here to *think like a scientist*. That's what the Science and Engineering Practices (SEPs) are all about. They represent the key skills that real scientists and engineers use every day to investigate the world, solve problems, and build new technologies.

On the Regents Exam in Life Science: Biology, you'll be expected to use these skills—not just recognize them. That means interpreting data, designing experiments, modeling biological systems, and explaining scientific ideas clearly. The exam is designed to mirror how science works in the real world.

Let's take a closer look at each SEP and how it shows up in your biology studies—and on the test.

1. Asking Questions and Defining Problems

What it means: Science starts with curiosity. Scientists ask questions about natural phenomena, while engineers define problems that need solutions.

In biology class: You might ask: What causes leaves to change color in autumn? or Why do some bacteria resist antibiotics?

On the test: You may be given a scenario and asked to identify a testable question or propose a scientific investigation.

Example Question: "Based on the observed increase in algae growth in a lake, what question could a scientist ask to investigate the cause?"

2. Planning and Carrying Out Investigations

What it means: Scientists design experiments to test hypotheses. They choose variables, control conditions, collect data, and ensure reliability.

In biology class: You might conduct a lab to see how temperature affects enzyme activity or how different soils affect plant growth.

On the test: You could be asked to identify independent and dependent variables, describe controls, or analyze a proposed experimental design.

Example Question: "Design an investigation to test how light intensity affects the rate of photosynthesis in aquatic plants."

3. Analyzing and Interpreting Data

What it means: Scientists use graphs, tables, and charts to make sense of what they observe. Analyzing data helps uncover patterns and draw conclusions.

In biology class: You might look at a graph showing changes in population size or enzyme activity over time.

On the test: You'll be asked to interpret visual data, find trends, and explain what the data show.

Example Question: "Use the data table to determine during which temperature range the enzyme activity is highest."

4. Constructing Explanations and Designing Solutions

What it means: Scientists explain *why* things happen based on evidence. Engineers use what they know to design solutions to real-world problems.

In biology class: You may explain how the human respiratory system functions or how natural selection explains changes in a species over time.

On the test: You might be asked to write a short explanation using scientific reasoning or propose a solution to a biological problem.

Example Question: "Construct a scientific explanation for how bacteria develop resistance to antibiotics."

5. Using Mathematics and Computational Thinking

What it means: Scientists use math to measure, calculate, and analyze data. Computational thinking helps with modeling systems or solving complex problems.

In biology class: You might calculate genetic probabilities with Punnett squares to determine population density.

On the test: You could be asked to solve problems, interpret numerical data, or use ratios and percentages.

Example Question: "Use a Punnett square to determine the probability that offspring will inherit a recessive trait."

6. Engaging in Argument from Evidence

What it means: Science involves discussion and debate. Scientists support their ideas with **evidence**, and challenge ideas that aren't supported by data.

In biology class: You might argue for or against a claim using evidence from a lab or an article.

On the test: You'll be asked to evaluate claims, identify supporting evidence, or construct an argument of your own.

Example Question: "Based on the data from the experiment, argue whether increased CO_2 levels are directly linked to rising ocean temperatures."

7. Developing and Using Models

What it means: Models are tools scientists use to represent systems, explain processes, and make predictions. Models can be physical, visual, or conceptual.

In biology class: You may draw a model of a DNA molecule or use a diagram to show nutrient cycling in an ecosystem.

On the test: You'll be asked to interpret, use, or even modify a model to explain a biological concept.

Example Question: "Use the diagram of the circulatory system to trace the path of oxygenated blood from the lungs to the body."

Why SEPs Matter

Every question on the 2025 Regents Biology Exam is designed to reflect *real science*. This means you won't just be recalling information—you'll be *doing* something with it. You will:

- Analyze experiments
- Interpret graphs and diagrams
- Apply knowledge to new situations
- Use models and construct explanations
- Justify your answers with data

How to Practice SEPs

To build your skills:

- Ask your own questions during lessons and labs
- Think critically when you read science articles or look at data
- Explain your reasoning out loud or in writing
- Use visuals and models to study and summarize big ideas
- Get hands-on with labs or simulations whenever possible

Final Thought: Practice Makes a Scientist

The more you practice using these skills, the more confident you'll be—not just for the Regents Exam, but for understanding and exploring the living world. Science is something you *do*—and with these practices, you're well on your way.

Disciplinary Core Ideas (DCIs)

The Disciplinary Core Ideas (DCIs) are the *heart and soul* of what you'll study in biology. These are the core content areas—the essential topics that scientists agree every biology student should understand by the end of high school.

When you take the Regents Exam in Life Science: Biology, you'll be tested on these major ideas. The questions won't just ask for facts—they'll ask you to apply these ideas, use data, analyze models, and explain phenomena.

Each DCI focuses on a different level of biological organization, from the tiniest molecules to the biggest ecosystems. Let's break them down:

LS1: From Molecules to Organisms—Structure and Processes

What it's about: This core idea focuses on how living things are built (structure) and how they stay alive (processes). It covers everything from the molecular building blocks of life to how entire body systems work.

What you'll need to know:

- The structure and function of cells and organelles
- Cell division (mitosis and the cell cycle)
- Photosynthesis and cellular respiration
- Body systems and how they maintain homeostasis
- How living things grow, develop, and get energy

Example Question: "Compare the processes of photosynthesis and cellular respiration and explain how energy is transformed in each."

LS2: Ecosystems—Interactions, Energy, and Dynamics

What it's about: Life doesn't happen in a bubble. This DCI is all about how organisms interact with each other and with their environment. It also looks at how energy flows through ecosystems and how matter cycles.

What you'll need to know:

- Food chains and food webs
- Energy pyramids and trophic levels
- Ecosystem stability and biodiversity
- Human impact on the environment (pollution, climate change, etc.)
- Relationships between organisms (predation, competition, symbiosis)

Example Question: "Use a food web to identify producers, consumers, and decomposers, and explain how energy moves through the ecosystem."

LS3: Heredity—Inheritance and Variation of Traits

What it's about: This section is all about genetics—how traits are passed from one generation to the next and how variations arise. You'll learn how DNA determines traits and how organisms inherit genetic information.

What you'll need to know:

- The structure and role of DNA and genes
- Dominant and recessive alleles
- Punnett squares and probability
- Mutations and their effects on traits
- Genetic technologies (cloning, genetic engineering)

Example Question: "Using a Punnett square, predict the probability of offspring having a certain trait, and explain how the alleles from each parent contribute."

LS4: Biological Evolution—Unity and Diversity

What it's about: Why are there so many different species on Earth? Why do some species go extinct while others thrive? This section explores how life has changed over time and the evidence for evolution.

What you'll need to know:

- Natural selection and adaptation
- The role of genetic variation in evolution
- Common ancestry and homologous structures
- Fossil evidence and molecular comparisons
- Speciation and extinction

Example Question: "Explain how natural selection can lead to the evolution of a population over time using a real-world example (like antibiotic resistance in bacteria)."

Earth and Space Science Connections

Why it matters: Living things are affected by Earth's systems, including climate, geology, and weather patterns. Some Regents questions may connect biological concepts to these larger Earth processes.

What you might see:

- The carbon cycle and climate change
- How changes in Earth's atmosphere affect ecosystems
- Long-term environmental trends and evolution

Example Question: "Describe how an increase in global temperatures can affect the migration and survival of a species."

Engineering and Math Connections

Why it matters: Scientists use engineering practices to solve problems and math to measure, model, and analyze biological data. You'll sometimes be asked to calculate, graph, or design.

You might be asked to:

- Analyze a data table from a scientific experiment
- Design a model of a feedback mechanism (like blood sugar regulation)
- Use ratios or percentages in genetics problems

Example Question: "Design a model that shows how insulin helps regulate blood glucose levels and describe what happens when this system is disrupted."

How to Master the DCIs

To succeed on the Regents Exam:

- Focus on *understanding*, not just memorizing
- Practice applying concepts in new situations
- Use models, data sets, and real-world examples
- Study how the DCIs connect to each other (for example, how heredity relates to evolution)

Final Thought: The Big Picture

Each DCI represents a major branch of biology. Together, they help you make sense of the living world—how it works, how it changes, and how we can protect it. Whether you're studying how cells use energy, how species evolve, or how humans impact ecosystems, the Disciplinary Core Ideas give you the tools to think scientifically—and to ace the Regents Exam.

Crosscutting Concepts (CCCs)

Science isn't just about memorizing facts—it's about *thinking like a scientist.* That means being able to make connections across different topics and disciplines. That's where Crosscutting Concepts (CCCs) come in. These big-picture ideas are used by scientists in every field, and they help you see patterns and relationships in the world around you.

Each CCC is like a lens you can use to examine a scientific problem. On the Regents Biology Exam, you'll be expected to apply these concepts to real-world data, situations, and problems.

Let's explore each one:

1. Patterns

What it means: Scientists look for repeating patterns to make predictions and discover relationships. In biology, patterns can be found in DNA sequences, life cycles, population trends, and more.

On the test: You might be shown a series of images or data sets and asked to identify a pattern, such as the inheritance of traits or the spread of a disease.

Example Question: "Based on the pattern of inheritance in the pedigree chart, what is the most likely genotype of the parents?"

2. Cause and Effect

What it means: Every action has a consequence. Understanding the cause of a biological event helps scientists figure out *why* things happen and *what* might happen next.

On the test: You may have to explain how one variable affects another—for instance, how a lack of insulin causes symptoms of diabetes.

Example Question: "Describe the cause-and-effect relationship between increased atmospheric CO_2 and ocean acidification."

3. Scale, Proportion, and Quantity

What it means: Biological processes happen at many scales—from microscopic cells to entire ecosystems. Scientists use math to measure changes and compare systems.

On the test: You might need to analyze graphs or perform basic calculations involving size, time, or concentration.

Example Question: "Compare the relative size and function of mitochondria and chloroplasts in a plant cell."

4. Systems and System Models

What it means: Scientists often study complex systems by modeling how different parts interact. Systems could be the human body, a food web, or even a single cell.

On the test: You could be asked to create or interpret a model showing how the parts of a system work together. For example, you may be asked to explain how nutrients move through an ecosystem.

Example Question: "Use a model to show how energy flows from producers to consumers in a food chain."

5. Energy and Matter

What it means: Energy is needed for all biological processes, and matter (like nutrients or gases) cycles through living systems. Understanding these flows helps explain everything from metabolism to ecosystems.

On the test: You'll analyze how energy enters and moves through systems, or how matter is transformed and reused.

Example Question: "Explain how matter is conserved during photosynthesis and cellular respiration."

6. Structure and Function

What it means: The shape and design of something often tell you how it works. This is especially true in biology—think about how the wings of birds are structured for flight or how enzymes are shaped to fit specific molecules.

On the test: You'll be asked to connect the structure of a cell, an organ, or a protein to its function.

Example Question: "Describe how the shape of red blood cells helps them carry oxygen efficiently."

7. Stability and Change

What it means: Some systems remain stable over time, while others change quickly or gradually. Scientists study both the *mechanisms of stability* and the *causes of change*.

On the test: You may be asked to explain how organisms maintain homeostasis or how an ecosystem responds to a disturbance.

Example Question: "Explain how feedback mechanisms help the body maintain internal stability during exercise."

Putting It All Together

Often, more than one crosscutting concept will apply to a single question. For example, you might examine *patterns* in data while also describing a *cause-and-effect* relationship and considering *energy flow* in a system.

Tip: When practicing, ask yourself:

- What pattern do I see?
- What is causing this to happen?
- What is the role of structure or energy here?
- How does this system work as a whole?

Being able to think in these ways—not just memorize content—is key to acing the new Regents Exam.

Chapter 3
Investigations—Learning Science by Doing Science

As a student preparing for the Regents Examination in Life Science: Biology, you're probably used to studying facts, reading explanations, and answering practice questions. But did you know that doing science—through hands-on investigations—is just as important as reading about it?

This chapter introduces you to a new and exciting part of science learning in New York State: Investigations. These aren't quizzes or multiple-choice tests. They're performance-based science tasks that give you the chance to think, work, and reason like a real scientist or engineer.

Let's explore what Investigations are, how they help you build important skills, and what they look like in your classroom.

What Are Investigations?

Real Science, Not Just a Lab Report

Investigations are hands-on, minds-on science activities that are now part of the high school biology experience across New York State. These tasks are meant to be **authentic**, meaning they resemble the way scientists and engineers actually solve problems in the real world.

They are:

- **Curriculum-embedded**—done in class as part of your learning
- **Locally administered**—created and scored by your teachers or district
- **Performance-based**—you show what you know by doing, not just writing
- **Aligned to the state's science standards**—helping you prepare for the Regents exam

Note: Investigations are **not part of the Regents exam itself**, but they help you develop the same kinds of **scientific thinking and reasoning skills** that the Regents will assess.

Why Are Investigations Important?

Think of Investigations as **practice for real-world science**. You get to explore biological phenomena, ask questions, test ideas, analyze data, and communicate your findings—just like a professional biologist.

How Investigations Help You:

- Reinforce the three dimensions of science learning:
  - Science and Engineering Practices (SEPs)—what scientists do
  - Disciplinary Core Ideas (DCIs)—what scientists know
  - Crosscutting Concepts (CCCs)—how ideas connect across science
- Develop your ability to:
  - Plan and carry out experiments
  - Build and use models
  - Interpret graphs and tables
  - Support your explanations with data
  - Think critically and reason through scientific problems

These are the exact **skills** you'll need for the **Regents clusters**, especially the **constructed-response** questions where you'll be asked to make claims, analyze patterns, or evaluate models.

What Does an Investigation Look Like?

Let's walk through a few example scenarios so you know what to expect.

Example 1: Modeling Enzyme Activity

Scenario: You're given a model of how temperature affects the shape of an enzyme and asked to predict how enzyme activity changes across a temperature range.

What you might do:

- Use a simulation or graph to identify the **optimal temperature**
- Explain what happens to the **active site** at higher temperatures
- Compare predictions to real lab data from enzyme reactions

Skills used: Developing and using models | Constructing explanations | Analyzing data

How it helps: This mirrors the kinds of data interpretation questions you'll see on clusters such as "Structure and Function."

Example 2: Investigating Photosynthesis and Light

Scenario: You place aquatic plants under different light intensities and measure how many oxygen bubbles they produce.

What you might do:

- Design a fair test by controlling variables like water temperature and CO_2
- Collect data on oxygen output
- Graph your results and explain the relationship between light and photosynthesis

Skills used: Planning and carrying out investigations | Using math and computational thinking

How it helps: Reinforces concepts from "Matter and Energy in Organisms and Ecosystems" and builds experience analyzing graphs—just like on the Regents.

Example 3: Ecosystem Modeling—Wolves in Yellowstone

Scenario: You're given a food web and population data from before and after wolves were reintroduced into Yellowstone National Park.

What you might do:

- Identify trophic interactions
- Use the data to explain cause-and-effect changes in the ecosystem
- Construct a model showing how biodiversity changed over time

Skills used: Interpreting models | Arguing from evidence | Cause-and-effect reasoning

How it helps: This connects directly to the type of reasoning used in ecosystem clusters on the Regents.

Investigations vs. Regents Clusters

Investigations	Regents Clusters
Hands-on tasks done in class	Written tasks done during the state exam
Locally scored and flexible in timing	Standardized and scored by the state
Focus on applying science and problem-solving	Focus on analyzing data and explaining reasoning
Real-time exploration and discussion	Independent written responses under timed conditions
Build the same skills tested on the Regents	Assess those skills through questions and CER tasks

Final Thoughts: Why This Matters to You

Investigations are more than just school assignments—they're your **training ground** for thinking scientifically. By participating in classroom investigations, you'll gain confidence in:

- Analyzing data
- Making sense of real-world problems
- Explaining biological phenomena clearly and with evidence

These experiences will make the Regents exam feel **familiar**, not intimidating. You'll be better prepared to tackle **data-rich clusters**, **models**, and **constructed-response questions** because you've already practiced those skills in a meaningful way.

So when your teacher says, "Today we're doing an Investigation," take it seriously—and give it your best. You're not just preparing for a class grade ... you're training to succeed on the Regents.

Chapter 4
The 2025 Regents Biology Exam—Format, Topics, and What to Expect

You've learned about the three dimensions of the New York State P–12 Science Learning Standards. Now it's time to get specific: What's actually on the Regents Examination in Life Science: Biology? How is the test set up? What topics are covered? And how much should you focus on each?

This chapter answers those questions and more—so you'll know exactly what to expect on exam day.

Test Format Overview

The new Regents Biology Exam is three hours long and will include both multiple-choice and constructed-response questions. Here's how it breaks down:

Section	Question Type	Estimated % of Exam
Multiple-Choice Questions	Choose from 4 answer choices	~60%
Constructed-Response Questions	Write your own answers (short answers, data analysis, models, etc.)	~40%

You'll answer 45–55 questions, grouped into 9 to 11 clusters (groups of related questions).

How Questions Are Organized: Clusters and Storylines

Unlike older exams, the new Regents Biology exam is built around **question clusters**. Each cluster:

- Is based on a real-world scientific phenomenon
- Includes multiple stimuli (e.g., graphs, charts, data sets, photos, passages)
- Asks related questions that build toward an explanation, model, or solution
- Includes both multiple-choice and constructed-response items

Think of each cluster as a mini science story. You'll analyze information, interpret data, and explain your thinking just like a real scientist.

Topic Breakdown: How Much of Each Topic Is on the Test?

The chart below shows the topic-level test blueprint from the NYSED Educator Guide. It tells you what percentage of the exam is focused on each topic area:

Topic	% of Questions
Structure and Function	9–15%
Matter and Energy in Organisms and Ecosystems	18–29%
Interdependent Relationships in Ecosystems	14–24%
Inheritance and Variation of Traits	14–24%
Natural Selection and Evolution	14–24%
Earth's Systems	5–11%
Engineering, Technology, and Applications of Science	3–11%

Types of Stimuli You'll See

Every question cluster will include **stimuli**, or real-world data and visuals, such as:

- Graphs and data tables
- Diagrams and scientific models
- Photos and real case studies
- Short reading passages

You'll use these materials to answer related questions that assess your ability to **interpret information, analyze patterns,** and **construct explanations**.

Question Types You'll Answer

You'll face two kinds of questions throughout the exam:

Multiple-Choice Questions

- Standard format: choose the best answer from four options
- Tests recall, interpretation, and basic application skills

Constructed-Response Questions

- Short answers that require writing
- May ask you to interpret data, draw conclusions, support claims with evidence, or develop models
- Designed to show how well you understand *and apply* what you've learned

Calculator Use

You may use a four-function or scientific calculator during the exam. Graphing calculators are NOT allowed.

Final Takeaways

- The 2025 Regents Biology exam is built around **real-world phenomena and scientific reasoning**, not just memorization.
- Expect a balanced mix of **multiple-choice** and **written responses**, with questions organized into **story-based clusters**.
- Practice using **stimuli**—like data tables, graphs, and models—to answer questions just as you'll do on the real test.

Chapter 5

Test-Taking Tips for the Regents Biology Exam

The Regents Biology exam is not just about memorizing facts—it's about applying your knowledge to **real-world science problems**, interpreting **data and visuals**, and **explaining your reasoning**. This chapter provides tips to help you feel calm, focused, and ready to tackle the exam with confidence.

Before the Test: Set Yourself Up for Success

1. Understand the Exam Format

- You'll answer questions in **clusters** (mini science scenarios).
- Each cluster includes **stimuli** (charts, models, or passages) and **9–11 related questions**.
- The exam includes:
  - **Multiple-choice** questions (about 60%)
  - **Constructed-response** questions (about 40%)
- Total test time: **3 hours**

2. Know the Three Dimensions

- Every question is based on:
  - A **Science and Engineering Practice (SEP)**—what scientists do
  - A **Disciplinary Core Idea (DCI)**—what scientists know
  - A **Crosscutting Concept (CCC)**—how science connects ideas
- Get familiar with what each one looks like in a question.

3. Study Strategically

- Review **each topic cluster** in this book.
- Practice explaining phenomena and models using scientific vocabulary.
- Create flashcards for **key vocabulary** and **processes** (e.g., photosynthesis, natural selection).
- Practice **reading graphs, tables, and models**—this is a skill you'll use on almost every question cluster.

During the Test: Smart Strategies for Every Question

1. Read the Stimulus Carefully

- Take your time reading the **phenomenon and accompanying data** before jumping to the questions.
- Underline or circle important information (e.g., trends in a graph, cause-effect words in a paragraph).

2. Don't Rush the Clusters

- Questions are **related** and often build in complexity.
- Use information from earlier questions to help with later ones.
  Example: If a graph shows rising CO_2 levels, a later question might ask why coral reefs are affected.

3. Eliminate Wrong Answers

For multiple-choice questions:

- Cross out clearly wrong answers.
- Look for **key words** like "always," "only," or "most likely"—these can help identify traps.
- If stuck, **make an educated guess**—there's no penalty for guessing.

4. Master Constructed Responses

- Use the **CER (Claim-Evidence-Reasoning)** method when appropriate:
  - **Claim**—a direct answer to the question
  - **Evidence**—data or information from the stimulus
  - **Reasoning**—the science behind why the evidence supports the claim
- Use **scientific vocabulary** (e.g., allele, feedback, carbon cycle).
- Write in **complete sentences**. Avoid "I think" or "I believe"—be confident in your science!

5. Manage Your Time

- Don't get stuck for too long on any one question.
- Aim to spend **no more than 15–20 minutes per cluster**.
- If a question seems difficult, **skip it and come back later**.

Bonus Tips for High-Scoring Answers

- **Use models and data** in your explanation. Refer directly to graphs or tables ("As shown in Figure 2 . . .").
- **Label diagrams** if asked to draw one. Use arrows, keywords, and clear titles.

- **Review your writing** and check that your answers match what the question asked—especially for multi-part questions.

Calm Your Nerves: Test Day Checklist

- Get a good night's sleep.
- Eat a healthy breakfast.
- Bring a pen, a pencil, and a permitted calculator.
- Breathe deeply and stay focused.
- Use **positive self-talk**: "I've practiced this. I know how to read models. I've got this!"

Final Advice

Remember, this test isn't about memorizing every science fact—it's about **thinking like a scientist**. That means:

- Reading carefully
- Explaining your thinking
- Applying knowledge to new situations
- Looking for patterns and evidence

The Regents Biology exam is your chance to show how much you've grown as a science learner. With the right strategies, mindset, and preparation, you can excel.

Good luck—you're ready!

Chapter 6

Regents-Style Practice Clusters

What's This Chapter About?

This chapter features original question clusters designed specifically to mirror the format of the Regents Examination in Life Science: Biology. These clusters are based on the New York State P–12 Science Learning Standards and use a three-dimensional approach that combines:

- Science and Engineering Practices (SEPs)—what scientists do
- Disciplinary Core Ideas (DCIs)—the big ideas in biology
- Crosscutting Concepts (CCCs)—themes that connect all sciences

Rather than testing isolated facts, the Regents exam asks you to **analyze real-world scientific situations** using models, data, graphs, and case studies.

How to Use This Chapter

You'll find one complete question cluster for each of the 7 major topic areas tested on the exam:

1. Structure and Function
2. Matter and Energy in Organisms and Ecosystems
3. Interdependent Relationships in Ecosystems
4. Inheritance and Variation of Traits
5. Natural Selection and Evolution
6. Earth's Systems
7. Engineering, Technology, and Applications of Science

Each cluster includes:

- A real-world **phenomenon**
- **9–11 questions** (just like the actual exam)
- Detailed **answer explanations** at the end of each cluster

Your Goal

Use these clusters to sharpen your skills in thinking, reasoning, and explaining like a biologist. As you work through each one:

- Read the phenomenon carefully
- Analyze the diagrams and data
- Use evidence to explain your reasoning
- Check your answers and learn from mistakes

Topic 1: Structure and Function

Science and Engineering Practice (SEP): Constructing Explanations and Designing Solutions

Disciplinary Core Idea (DCI): LS1.A: Structure and Function

Crosscutting Concept (CCC): Structure and Function

Phenomenon: Enzymatic Activity in Digestion and Feedback Regulation in the Human Body

Base your answers to questions 1 through 9 on the information below and on your knowledge of biology.

Investigating Enzymes in the Digestive System

Enzymes are proteins that speed up chemical reactions in the body. Each enzyme has a unique three-dimensional shape that determines its function. The human digestive system relies on enzymes to break down food into nutrients. These enzymes work best at specific temperatures and pH levels.

A team of scientists studied an enzyme called lactase, which helps digest lactose (a sugar found in milk). They measured lactase activity at different temperatures.

Table 1. Lactase Activity at Different Temperatures

Temperature (°C)	Lactase Activity (Arbitrary Units)
10	2
25	5
37	9
45	4
60	1

1. What conclusion is best supported by the data in Table 1?

(1) Lactase activity increases indefinitely as temperature increases.

(2) Lactase functions best at low temperatures.

(3) Lactase is most active near normal human body temperature.

(4) Lactase becomes more efficient after it is denatured.

2. Using evidence from Table 1, explain why lactase activity drops after 45°C. Include how temperature affects enzyme structure and function.

____

____

____

____

Enzyme Structure and Specificity

The function of enzymes depends on their shape. High temperatures or extreme pH can change an enzyme's shape, causing denaturation. Denatured enzymes cannot bind to their substrates.

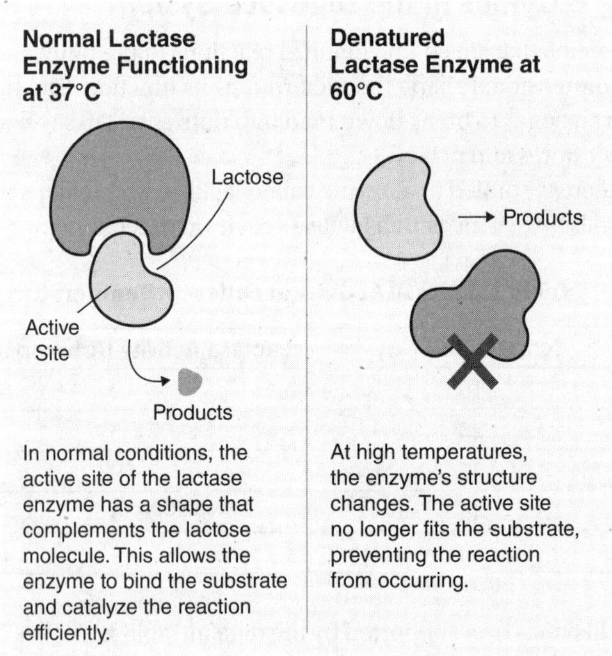

3. What most likely explains the drop in lactase activity at 60°C?

    (1) The substrate was no longer present.

    (2) The enzyme was denatured and lost its shape.

    (3) Lactase was consumed during the reaction.

    (4) Lower temperatures caused the enzyme to speed up.

4. Using the model above, describe how enzyme-substrate interaction is affected by denaturation. Explain why this prevents the breakdown of lactose.

____

____

____

____

Homeostasis and Feedback

After digestion, glucose enters the bloodstream. The hormone insulin helps cells absorb glucose. This is part of a feedback mechanism that maintains homeostasis.

Table 2. Blood Glucose Levels Before and After Eating

Time (minutes)	Glucose Level (mg/dL)
0	90
30	140
60	120
90	100
120	90

5. Which process is most responsible for returning glucose levels to baseline after eating?

(1) Digestive enzyme activity
(2) Cell division in the pancreas
(3) Positive feedback from glucagon
(4) Negative feedback regulated by insulin

6. Use Table 2 to explain how the body uses feedback to maintain stable blood glucose levels after a meal. Include the roles of insulin and target cells.

____

____

____

____

A Genetic Mutation in an Enzyme

A student reads about a genetic mutation that changes a single amino acid in the lactase enzyme. The mutated version is less efficient at breaking down lactose.

7. Which statement best explains how a change in one amino acid can reduce enzyme function?

(1) The enzyme is entirely replaced with a different protein.
(2) The enzyme becomes larger and stronger.
(3) The change alters the enzyme's shape, affecting the active site.
(4) The mutation makes the enzyme reproduce faster.

8. Explain how a mutation that changes the shape of lactase affects its function. Predict how this mutation could impact an individual's ability to digest milk.

____

____

____

____

9. **Claim:** "The structure of a biological molecule determines its function."

Using examples from the cluster (enzymes, feedback, or hormones), construct a well-supported argument to explain how structure affects function in biological systems.

____

____

____

____

Answer Explanations

1. **(3)** The data table shows that lactase activity is highest at 37°C, which is the average internal temperature of the human body. This suggests that the enzyme has evolved to function optimally in that environment. The optimal temperature for enzyme activity is the temperature at which the enzyme's shape (especially its active site) is ideal for catalyzing its reaction efficiently. At this temperature, the enzyme maintains proper tertiary structure—its 3D folding—allowing it to bind the substrate (lactose) and perform its function. At lower or higher temperatures, this shape begins to destabilize, reducing activity.
2. **Sample Response:** Lactase shows the highest rate of activity at 37°C, indicating this is its optimal temperature. As the temperature increases to 60°C, the enzyme's activity sharply drops. This is due to a process called denaturation, where the enzyme's protein structure unravels, especially the active site—the part that fits with the substrate. When this shape is lost, the enzyme can no longer bind lactose effectively, and the reaction slows down or stops entirely. This shows that enzyme structure is directly linked to function—any change to the shape can reduce or eliminate its effectiveness.
3. **(2)** High temperatures can cause proteins, like enzymes, to denature, which means they lose their specific shape. Enzymes rely on the exact structure of their active site to bind with substrates. When the enzyme is exposed to excessive heat, the bonds that hold its shape—especially hydrogen bonds—are disrupted. This changes the active site and prevents proper binding. Thus, the sharp drop in enzyme activity at 60°C is most likely due to denaturation rather than to substrate absence or enzyme consumption.
4. **Sample Response:** Panel B shows a denatured enzyme whose active site no longer matches the shape of the lactose molecule. In normal function (Panel A), the active site fits the substrate like a "lock and key." But when the enzyme is denatured—often by high heat or pH changes—the structure is altered, and the substrate cannot bind. Without this interaction, the enzyme cannot catalyze the breakdown of lactose into glucose and galactose. This illustrates the principle that the shape of a protein determines its biological function.
5. **(4)** Insulin is a hormone released by the pancreas in response to rising blood glucose levels after eating. It signals cells to take in glucose, reducing the amount in the bloodstream. This is an example of negative feedback, a regulatory mechanism where the body responds to a stimulus by counteracting the change. Once glucose levels return to normal, insulin release slows. This feedback loop helps maintain homeostasis, or internal balance. Option (4) best describes this biological process.

6. **Sample Response:** After a meal, glucose levels in the blood increase. The pancreas detects this change and releases insulin, which binds to receptors on body cells, signaling them to take in glucose for energy or storage. As glucose is removed from the blood, the concentration decreases, returning to normal levels. This is a feedback loop that maintains homeostasis—the stable internal environment necessary for proper cell function. Without insulin or this feedback, blood sugar would stay dangerously high, as in people with diabetes.
7. **(3)** Proteins, like enzymes, are built from chains of amino acids. The sequence and chemical properties of these amino acids determine the final 3D shape of the protein. A mutation that changes even one amino acid can alter the folding pattern and shape of the active site. This prevents proper binding to the substrate. In the case of lactase, a single change could result in reduced efficiency or even total loss of function—leading to conditions like lactose intolerance.
8. **Sample Response:** The lactase enzyme has a specific shape that fits the lactose molecule. If a genetic mutation changes even one amino acid, the shape of the enzyme's active site may also change. This prevents lactose from binding properly. As a result, lactose remains undigested in the digestive tract, leading to symptoms like bloating and cramping—this is known as lactose intolerance. This illustrates how genetic changes can impact protein structure and, in turn, organism function.
9. **Sample Response:** In biology, structure determines function. Enzymes must have the correct shape to interact with specific substrates. If the structure changes, such as through denaturation or mutation, their function is lost. Similarly, hormone receptors have shapes that match specific hormones. For example, insulin must bind to a receptor with a complementary shape to signal glucose uptake. If the receptor's shape changes (due to mutation), it may no longer work. These examples demonstrate how vital protein structure is for maintaining biological functions like digestion, feedback loops, and cell communication.

Topic 2: Matter and Energy in Organisms and Ecosystems

Science and Engineering Practice (SEP): Developing and Using Models

Disciplinary Core Idea (DCI): LS1.C: Organization for Matter and Energy Flow in Organisms

Crosscutting Concept (CCC): Energy and Matter—Flows, Cycles, and Conservation

Phenomenon: Energy Flow in a Seagrass Ecosystem and the Effects of Algal Blooms

Base your answers to questions 1 through 10 on the information below and on your knowledge of biology.

Seagrass Ecosystem

Seagrass beds are aquatic ecosystems found in shallow coastal waters. These plants perform photosynthesis and form the base of the food web. Seagrass supports snails, shrimp, crabs, small fish, and predators like sea turtles and birds. It also stores large amounts of carbon.

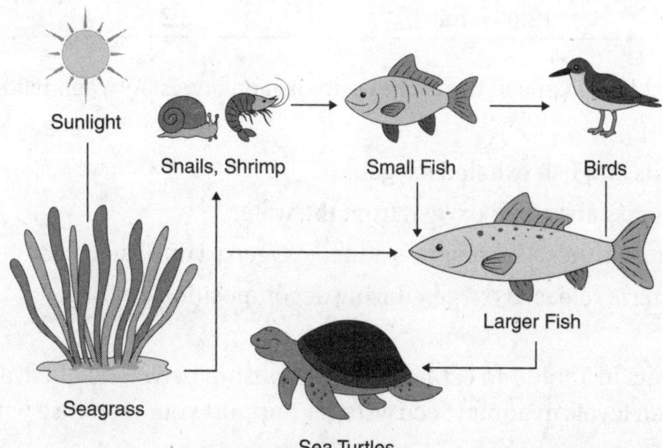

Figure 1. Food Web in a Seagrass Ecosystem

1. Which statement best explains how energy moves through the seagrass ecosystem?
   (1) Energy is recycled through all organisms and returned to the producers.
   (2) Energy is transferred from one trophic level to the next and lost as heat.
   (3) All energy is stored in the top predators and reused during decomposition.
   (4) Energy flows backward from the birds to the seagrass to complete the cycle.

2. Using Figure 1, describe how matter and energy flow from the Sun through the seagrass ecosystem. Include the role of producers, consumers, and decomposers in your explanation.

____

____

____

____

Oxygen Levels and Photosynthesis

Researchers monitored dissolved oxygen (DO) levels in a coastal bay with abundant seagrass.

Table 1. Dissolved Oxygen in the Water Over 24 Hours

Time (hour)	Dissolved Oxygen (mg/L)
6:00 a.m.	3.5
12:00 p.m.	7.8
6:00 p.m.	6.9
12:00 a.m.	2.2

3. What most likely caused the increase in oxygen levels between 6:00 a.m. and 12:00 p.m.?
   (1) Snails and fish exhaled oxygen.
   (2) Seagrass absorbed oxygen from the water.
   (3) Photosynthesis by seagrass added oxygen to the water.
   (4) Bacteria released oxygen during decomposition.

4. Use the data in Table 1 to explain the relationship between light availability and oxygen levels in aquatic ecosystems. Support your response with biological reasoning.

____

____

____

____

Algal Blooms and Ecosystem Disruption

In recent years, nutrient runoff from nearby farmland has caused seasonal algal blooms. The algae multiply quickly, blocking sunlight and eventually decomposing after they die.

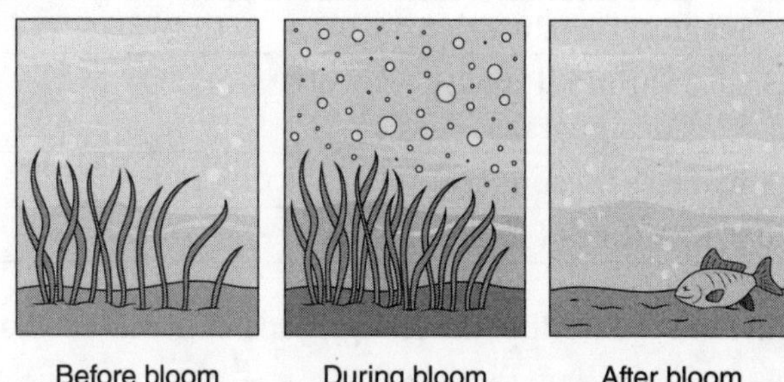

Figure 2. Impact of Algal Bloom on the Ecosystem

5. What is the most direct result of excessive algae growth in the bay?

(1) Fish and shrimp use algae as a new food source.

(2) Algae increase oxygen through photosynthesis at night.

(3) Decomposing algae reduce oxygen, harming aquatic animals.

(4) Algae absorb nitrogen, improving water quality.

6. Explain how an algal bloom disrupts the cycling of matter and flow of energy in the ecosystem. Include its effects on producers, consumers, and oxygen levels.

___

___

___

___

Energy Pyramid Data

Researchers estimated energy flow through the seagrass ecosystem.

Table 2. Energy Available at Each Trophic Level

Trophic Level	Energy (kcal/m²/year)
Seagrass (Producer)	10,000
Snails/Shrimp (Primary Consumers)	1,000
Small Fish (Secondary)	100
Large Fish/Birds (Tertiary)	10

7. Which statement best describes the energy transfer in this ecosystem?

(1) Most energy is stored at higher levels for predators.
(2) Only about 10% of energy is transferred from one level to the next.
(3) Energy is lost at each level because organisms do not eat enough.
(4) Energy is destroyed as it moves through the ecosystem.

8. Use the data in Table 2 to explain why energy availability limits the number of top-level predators in the ecosystem.

____

____

____

____

9. Which molecule provides the stored energy that producers use to build organic compounds?

(1) ATP
(2) CO_2
(3) Glucose
(4) H_2O

10. **Claim:** "Matter cycles but energy flows through ecosystems."

Using evidence from this cluster, support or refute this claim. Include examples of how carbon or oxygen cycle, and how energy availability changes at each level.

Answer Explanations

1. **(2)** In ecosystems, energy flows in one direction—from producers to consumers—and is lost as heat at each trophic level. Only a small fraction of energy (roughly 10%) is transferred to the next level; the rest is used for metabolic processes or lost to the environment. This principle explains why energy pyramids are wide at the base and narrow at the top. The energy that reaches the top-level consumers is significantly reduced compared to what producers initially absorb from sunlight.
2. **Sample Response:** Energy from the Sun is captured by seagrass, a producer, through photosynthesis. Seagrass converts light energy into chemical energy stored in glucose, which becomes available to primary consumers like snails and shrimp. As those animals are eaten by fish, energy moves up the food chain. However, not all energy is transferred—some is used for life processes, and much is lost as heat. When organisms die, decomposers break them down and recycle matter (like nutrients), but energy is not recycled—it exits the system as heat.
3. **(3)** Photosynthesis is a process in which aquatic plants and algae use sunlight to convert carbon dioxide and water into glucose and oxygen. During the daytime, when light is available, seagrass increases photosynthetic activity, releasing oxygen into the water. This explains the observed rise in dissolved oxygen during daylight hours, especially between 6:00 a.m. and 12:00 p.m. It's the most direct biological cause of the increase in oxygen levels in this time window.
4. **Sample Response:** Photosynthesis depends on light. As the Sun rises, light becomes more available, and aquatic producers like seagrass increase their photosynthetic rate. This produces more oxygen, which dissolves into the water, increasing the dissolved oxygen levels. At night, in the absence of light, photosynthesis stops, and respiration by plants and animals continues, consuming oxygen instead of producing it. This explains the daily fluctuation in oxygen levels in aquatic systems.
5. **(3)** Algal blooms occur when nutrient levels (often from fertilizer runoff) allow excessive algae growth. These algae block sunlight, harming submerged plants. When the algae die, decomposers (like bacteria) break them down, a process that uses large amounts of oxygen. This leads to hypoxia (low oxygen levels), which can kill fish and other animals. The question asks about the most direct result, and it is the oxygen depletion from decomposition, not the presence of algae itself.

6. **Sample Response:** An algal bloom creates a dense layer of algae on the water surface, blocking sunlight from reaching underwater plants like seagrass. Without light, these producers cannot photosynthesize and begin to die off. As dead algae and seagrass accumulate, decomposers break them down, consuming oxygen in the process. This reduces the available oxygen in the water, leading to fish kills and a breakdown of the food web. Both producers and consumers are negatively impacted, reducing ecosystem stability.
7. **(2)** The 10% rule is a general guideline in ecology stating that only about 10% of the energy from one trophic level is transferred to the next. The rest is used for growth, reproduction, or movement, or is lost as heat. This pattern explains the shape of energy pyramids, where top predators are few and require a large base of energy-rich producers and primary consumers to sustain them.
8. **Sample Response:** According to the data, the energy available at the top level of the pyramid is only 10 kcal, compared to 10,000 kcal at the producer level. This drastic decrease in energy means that only a small number of top predators can be supported. These predators need a large supply of prey to meet their energy demands. Ecosystems cannot support many predators because energy becomes increasingly scarce at higher levels, which is why most energy pyramids have a broad base and a narrow top.
9. **(3)** During photosynthesis, producers like seagrass create glucose, a sugar molecule that stores chemical energy. This energy is used by the plant for growth and is passed on to herbivores when they eat the plant. Glucose is the primary product of photosynthesis and the main source of energy at the base of the food web. While ATP is used for energy *within* cells, glucose is the energy currency that moves through ecosystems.
10. **Sample Response:** In ecosystems, matter cycles and energy flows. Elements like carbon, nitrogen, and oxygen are recycled through the biosphere, atmosphere, geosphere, and hydrosphere. For example, CO_2 is absorbed by seagrass for photosynthesis, then released back into the atmosphere during respiration or decomposition. Energy, on the other hand, enters through sunlight, is converted into glucose, and is passed from organism to organism. But it is eventually lost as heat—unlike matter, energy does not cycle. This one-way flow limits the size and complexity of higher trophic levels.

Topic 3: Interdependent Relationships in Ecosystems

Science and Engineering Practice (SEP): Using Mathematics and Computational Thinking

Disciplinary Core Idea (DCI): LS2.A: Interdependent Relationships in Ecosystems

Crosscutting Concept (CCC): Stability and Change

Phenomenon: Gray Wolf Reintroduction and Ecosystem Stability in Yellowstone

Base your answers to questions 1 through 9 on the information below and on your knowledge of biology.

Wolves Return to Yellowstone

In 1995, gray wolves were reintroduced to Yellowstone National Park after being absent for decades. Their absence had led to a dramatic rise in the elk population, which overgrazed young trees and vegetation, disrupting the park's ecosystems.

Over time, scientists observed changes in the populations of many species, including willows, beavers, coyotes, and birds.

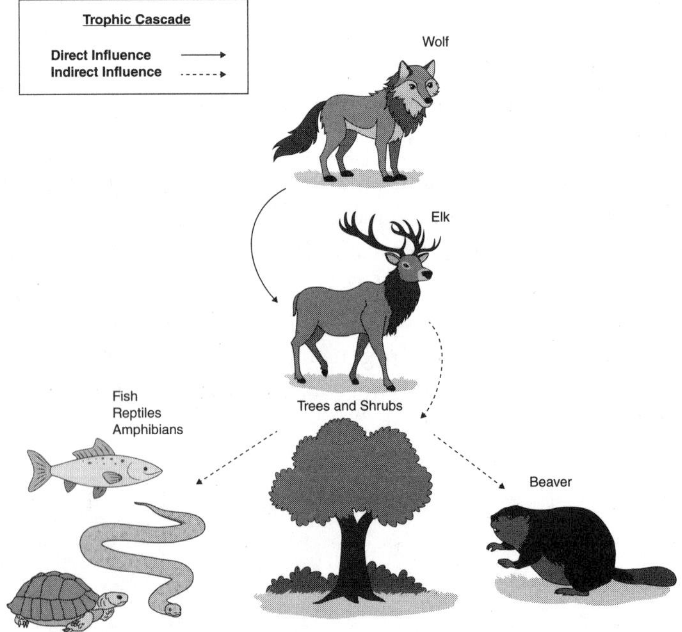

Figure 1. Simplified Trophic Interactions

1. How did the reintroduction of wolves most directly affect Yellowstone's food web?

(1) Increased predation on beavers improved wetland habitats.
(2) Wolves eliminated elk, causing a loss of herbivores.
(3) Wolves reduced elk numbers, allowing willows to recover.
(4) The presence of wolves directly reduced the beaver population.

2. Use evidence from Figure 1 to explain how the presence of wolves indirectly contributed to increased biodiversity in Yellowstone.

____

____

____

____

Carrying Capacity and Limiting Factors

Scientists tracked elk population data after the reintroduction of wolves.

Table 1. Elk Population and Wolf Population Over Time

Year	Elk Population	Wolf Population
1995	17,000	14
2000	12,000	100
2005	7,500	150
2010	6,200	140
2015	5,800	120

3. What do the data in Table 1 suggest about the relationship between wolves and elk?

(1) The elk population increased as the wolf population increased.
(2) Wolves are a limiting factor that contributed to the elk's population decrease.
(3) Wolves helped elk reach their carrying capacity.
(4) The wolf population decreased due to elk overgrazing.

4. Using data from Table 1, explain how the concept of carrying capacity applies to the elk population. What other factors besides wolves might affect elk numbers?

____

____

____

____

Changes in Vegetation and Water Systems

Willow trees are a key food source for beavers, which create wetlands that provide habitats for birds, amphibians, and fish. With fewer elk eating young willows, the trees began to recover. This led to new beaver colonies and changes in stream flow.

Table 2. Number of Beaver Colonies vs. Willow Density

Year	Avg. Willow Height (cm)	Number of Beaver Colonies
1995	35	1
2000	65	3
2010	120	8
2015	150	12

5. What is the most likely explanation for the trend shown in Table 2?
   (1) Increased elk grazing promoted willow growth.
   (2) More beavers caused willows to grow taller.
   (3) Fewer elk allowed willows to recover, supporting more beavers.
   (4) Willow trees competed with beavers for space and nutrients.

6. Describe the cause-and-effect relationships among elk, willows, and beavers. Use Table 2 to support your explanation.

____

____

____

____

Ecological Stability and Human Influence

In 2020, a proposal was made to remove wolves again due to livestock conflicts in nearby ranching areas. Ecologists warned that this could destabilize the ecosystem again.

7. Which outcome is most likely if wolves were removed again?

(1) Elk would decrease, and willows would continue to grow.
(2) Biodiversity would remain stable because of beavers.
(3) Elk numbers would increase, reducing willow and beaver populations.
(4) Willow trees would continue to increase due to higher sunlight.

8. Evaluate the potential ecological trade-offs of removing wolves again. Consider how the change might affect the stability of Yellowstone's ecosystems.

____

____

____

____

9. **Claim:** "The presence of a top predator like the gray wolf helps regulate ecosystem stability."

Using information from this cluster, construct a claim-evidence-reasoning response that supports or refutes this statement.

____

____

____

____

Answer Explanations

1. **(3)** The return of gray wolves to Yellowstone reduced the elk population, which had grown too large in the wolves' absence. Elk had been overgrazing young willows and other vegetation, preventing plant recovery and damaging habitats. With fewer elk, willow plants were able to regrow, providing food and shelter for other species such as beavers and songbirds. This demonstrates a trophic cascade, where the actions of a top predator indirectly affect multiple levels of an ecosystem.
2. **Sample Response:** The wolves' presence caused a decline in elk, their primary prey. As the number of elk decreased, grazing pressure on willows also decreased. This allowed the willows to grow taller and denser. The increased willow growth provided both building material and food for beavers, whose numbers increased. Beavers, in turn, created wetlands by damming streams, which formed habitats for amphibians, insects, birds, and fish. This chain of effects demonstrates how a top predator can increase biodiversity by restoring balance to an ecosystem.
3. **(2)** The graph showed an inverse relationship between wolf and elk populations: as wolf numbers increased, elk numbers decreased. This suggests that wolves acted as a limiting factor, controlling the elk population through predation. A limiting factor is any biotic or abiotic element that restricts population size. Without wolves, elk had few natural predators and their population grew unchecked, leading to ecosystem imbalance. The reintroduction of wolves reestablished that natural check.
4. **Sample Response:** At first, the elk population declined significantly as the number of wolves rose. Eventually, the elk population stabilized, suggesting it reached a new carrying capacity—the maximum number of individuals the environment can support without degrading habitat. Factors that affect carrying capacity include predation, food availability (willow and other plants), water, disease, and competition with other herbivores. The presence of wolves shifted the balance of the ecosystem and helped bring elk numbers into alignment with available resources.
5. **(3)** As elk populations dropped, willow plants recovered. Willows are a key food and building source for beavers. As the density and height of willows increased, more beavers could survive and reproduce, increasing the number of beaver colonies. The data in Table 2 show a direct relationship: as willow height increased, the number of beaver colonies also increased, showing a strong link between plant recovery and species resurgence.

6. **Sample Response:** Elk feed heavily on young willows. When their population was high, they overgrazed the plants, preventing them from reaching maturity. When wolves returned, elk numbers dropped, and willow plants began to recover. The regrowth of willows provided critical habitat and resources for beavers. As beavers returned and built dams, they slowed water flow, creating wetlands. These wetlands are rich habitats that increase biodiversity by supporting many species—fish, birds, amphibians, and insects.

7. **(3)** If wolves were removed again, elk populations would likely rise due to the loss of their primary predator. With more elk, grazing on young willows would intensify, leading to plant loss. Fewer willows would mean less food and building material for beavers, leading to a decline in their numbers. Fewer beavers would result in fewer wetlands, decreasing habitat variety and leading to less biodiversity. This chain of events would destabilize the ecosystem, reversing much of the positive change from the wolf reintroduction.

8. **Sample Response:** Removing wolves would likely lead to an uncontrolled elk population, as happened before 1995. This could cause severe overgrazing, especially of willows and other riparian vegetation. The loss of plant life would impact beavers, birds, and aquatic species, reducing biodiversity and altering water systems. While ranchers might benefit in the short term from fewer wolves (reduced predation on livestock), the ecological cost could be high. Trade-offs like these are critical when managing wild populations and ecosystems.

9. **Sample Response:** The presence of wolves regulated elk populations, which in turn allowed willows to recover. That recovery supported more beavers, who improved water retention and created wetland habitats. These effects increased biodiversity and helped restore ecosystem balance. Without wolves, these trophic interactions would collapse. The evidence supports the claim that top predators like wolves play a key role in maintaining ecosystem stability by controlling populations and triggering positive ecological feedback loops.

Topic 4: Inheritance and Variation of Traits

Science and Engineering Practice (SEP): Analyzing and Interpreting Data

Disciplinary Core Ideas (DCIs): LS3.A: Inheritance of Traits | **LS3.B:** Variation of Traits

Crosscutting Concept (CCC): Cause and Effect

Phenomenon: Genetic Variation in Sickle Cell Trait and Malaria Resistance

Base your answers to questions 1 through 9 on the information below and on your knowledge of biology.

The Sickle Cell Trait and Malaria

Sickle cell disease is caused by a mutation in the gene for hemoglobin, a protein in red blood cells that carries oxygen. People with two copies of the mutated allele (HbS) have sickle cell disease. However, people with only one copy (HbA / HbS) are usually healthy and are more resistant to malaria, a deadly parasitic disease transmitted by mosquitoes.

Table 1. Genotypes and Phenotypes Related to Sickle Cell and Malaria

Genotype	Phenotype	Malaria Resistance
HbA / HbA	Normal hemoglobin	No
HbA / HbS	Carrier (no disease symptoms)	Yes
HbS / HbS	Sickle cell disease	Yes

1. Which best explains why the HbA / HbS genotype is more common in regions where malaria is widespread?
   (1) It causes sickle cell disease, which prevents all infections.
   (2) It increases oxygen flow, making individuals healthier.
   (3) It provides resistance to malaria without causing disease symptoms.
   (4) It allows individuals to transmit both malaria and sickle cell disease.

2. Using Table 1, explain why natural selection has maintained the HbA / HbS genotype in certain populations.

____

____

____

____

Patterns of Inheritance

A Punnett square is used to model how two carrier parents (HbA / HbS) can pass on alleles to their children.

	HbA	HbS
HbA	HbA / HbA	HbA / HbS
HbS	HbA / HbS	HbS / HbS

Figure 1. Punnett Square—Carrier × Carrier Cross

3. What is the probability that two carrier parents will have a child with sickle cell disease?

(1) 0%

(2) 25%

(3) 50%

(4) 75%

4. Use the Punnett square in Figure 1 to determine the expected ratio of offspring phenotypes. Explain your reasoning.

____

____

____

____

Mutation and Protein Function

The mutation that causes sickle cell disease changes one amino acid in the hemoglobin protein. This small change causes red blood cells to become rigid and crescent-shaped. These cells can block blood flow and carry less oxygen.

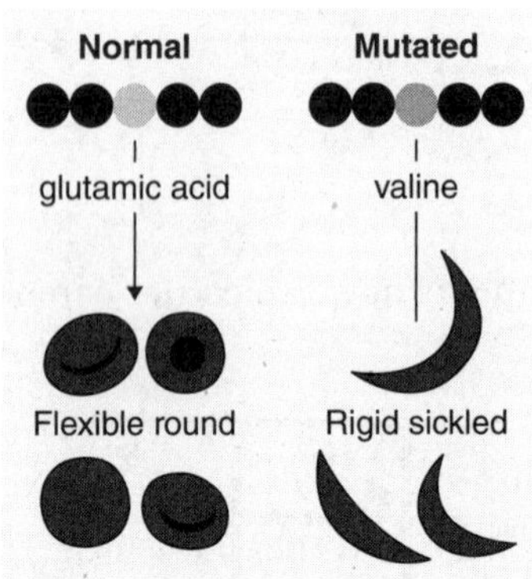

Figure 2. Comparison of Normal and Sickle Hemoglobin Proteins

5. What causes the structural change in red blood cells of individuals with sickle cell disease?

(1) A mutation in the DNA that alters protein shape and function
(2) A virus that destroys healthy hemoglobin
(3) The overproduction of red blood cells in the bone marrow
(4) Inheritance of the same recessive allele from both parents

6. Explain how a change in DNA leads to a change in protein shape and function in sickle cell disease. Include the relationship between genes, proteins, and traits.

____

____

____

____

Environmental Interactions with Genes

Researchers studied the frequency of the HbS allele in different African populations. The allele was more common in areas where malaria was prevalent and less common in high-altitude or desert regions where mosquitoes were rare.

Table 2. Frequency of HbS vs. Malaria Prevalence

Region	% Population with HbS Allele	Malaria Risk (Low–High)
West Africa	18%	High
Central Africa	15%	High
North Africa	3%	Low
South Africa	1%	Low

7. What do the data in Table 2 suggest about the relationship between environment and genetic traits?

(1) The HbS allele is randomly distributed in all environments.
(2) Environmental pressure from malaria selects for the HbS allele.
(3) People in high-malaria regions are immune to sickle cell disease.
(4) Climate is the only factor affecting allele frequency.

8. Describe how malaria acts as a selective pressure that influences allele frequency in human populations. Use Table 2 in your explanation.

____

____

____

____

9. Claim: "Variation in a population increases the chances that some individuals will survive environmental changes."

Using evidence from this cluster, construct an argument that supports this claim. Be sure to include how traits are inherited and why variation is important.

____

____

____

____

Answer Explanations

1. **(3)** Individuals with the heterozygous genotype HbA / HbS, commonly referred to as carriers, have one normal hemoglobin allele and one sickle cell allele. These individuals usually do not develop full-blown sickle cell disease but gain protection against malaria, a life-threatening disease caused by *Plasmodium* parasites transmitted by mosquitoes. This phenomenon is an example of heterozygote advantage, where being a carrier provides a survival benefit in malaria-endemic regions. As a result, natural selection favors this genotype in these environments, increasing its frequency in the population.
2. **Sample Response:** Carriers with the HbA / HbS genotype have red blood cells that are mostly normal but possess some resistance to malaria infection. This gives them a selective advantage in areas where malaria is common, such as parts of Africa, South Asia, and the Middle East. Individuals with the HbA / HbA genotype are susceptible to malaria, while those with HbS / HbS suffer from sickle cell disease, which can be debilitating or fatal. Therefore, natural selection favors the heterozygous genotype, leading to its persistence in regions where malaria is a major environmental pressure.
3. **(2)** The Punnett square in the question models the cross between two carriers (HbA / HbS). Each parent can pass on either the HbA or the HbS allele. The resulting genotypes are:

   HbA / HbA (normal)—25% chance

   HbA / HbS (carrier)—50% chance

   HbS / HbS (disease)—25% chance

   This means there is a 1 in 4, or 25% probability, that a child will inherit the HbS / HbS genotype and develop sickle cell disease.
4. **Sample Response:** The Punnett square shows four possible genotype combinations:

   1 out of 4: HbA / HbA → normal hemoglobin

   2 out of 4: HbA / HbS → carrier (resistant to malaria, no disease)

   1 out of 4: HbS / HbS → has sickle cell disease

   This gives a genotypic ratio of 1:2:1. However, when looking at phenotypes (what is physically expressed), three of the four offspring do not have sickle cell disease—one normal and two carriers. Therefore, the phenotypic ratio is 3 without disease : 1 with disease.

5. **(1)** A point mutation in the hemoglobin gene (specifically in the HBB gene on chromosome 11) causes the substitution of the amino acid valine for glutamic acid. This single amino acid change alters the structure of the hemoglobin protein, causing it to clump together under low oxygen conditions. The affected red blood cells become rigid and sickle-shaped, which limits their oxygen-carrying capacity and can block blood vessels, leading to pain and organ damage. This is a classic example of how a change at the DNA level leads to a change in protein shape and function and, ultimately, a change in phenotype.
6. **Sample Response:** Genes contain instructions for making proteins, and a mutation in a gene changes the sequence of DNA bases, which can alter the sequence of amino acids in the resulting protein. In the case of sickle cell disease, the substitution of just one base changes the protein structure, leading to misfolded hemoglobin. Misfolded hemoglobin causes red blood cells to form a sickle shape, which affects their ability to carry oxygen and flow through blood vessels. This is a clear example of how a genetic mutation → altered protein structure → visible trait (phenotype).
7. **(2)** Table 2 shows that populations in regions with high malaria risk, like West and Central Africa, have a much higher frequency of the HbS allele than populations in regions with low malaria risk. This indicates that natural selection has increased the frequency of the allele because it provides a survival advantage in the presence of malaria. This is a geographically linked example of evolution in action, where environmental factors directly influence allele frequency in a population.
8. **Sample Response:** In areas where malaria is prevalent, individuals with the HbA / HbS genotype have a higher survival rate. They are protected against malaria while avoiding the severe symptoms of sickle cell disease. This increases their reproductive success, allowing them to pass on the HbS allele to their offspring. Over generations, the frequency of the HbS allele increases in the population. This is an example of directional selection, where environmental pressures shift the genetic makeup of the population toward a trait that improves survival.
9. **Sample Response:** Within any population, genetic variation exists due to mutations and recombination. In environments with specific challenges, like malaria, some of that variation can lead to increased survival. In this case, carriers of the HbS allele are more likely to survive and reproduce in malaria-prone regions. Over time, this variation leads to a shift in the genetic makeup of the population—an example of evolution through natural selection. Genetic variation is essential because it provides the raw material on which selection acts, allowing populations to adapt to changing environments.

Topic 5: Natural Selection and Evolution

Science and Engineering Practice (SEP): Constructing Explanations and Designing Solutions

Disciplinary Core Ideas (DCIs): LS4.C: Adaptation | **LS4.B:** Natural Selection

Crosscutting Concept (CCC): Cause and Effect

Phenomenon: Antibiotic Resistance in Bacteria

Base your answers to questions 1 through 9 on the information below and on your knowledge of biology.

Bacterial Infections and Antibiotics

Antibiotics are medicines that kill bacteria or stop their growth. However, some bacteria have become resistant to certain antibiotics, meaning the medicine no longer works against them. This resistance happens when bacteria with certain genetic traits survive treatment and reproduce, passing on those traits.

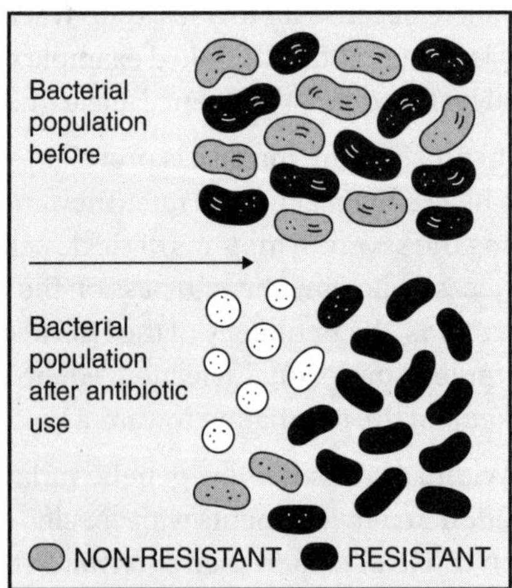

Figure 1. Bacterial Population Before and After Antibiotic Use

1. What best explains how antibiotic resistance increases in a bacterial population?
   (1) Bacteria learn to avoid antibiotics and pass that knowledge on.
   (2) Antibiotics cause bacteria to mutate in ways that make them resistant.
   (3) Only bacteria with resistance survive and reproduce after exposure.
   (4) Resistance traits are produced when antibiotics are not taken properly.

2. Use Figure 1 to explain how natural selection leads to an increase in antibiotic-resistant bacteria. Be sure to describe variation, selection, and reproduction.

____

____

____

____

Genetic Variation and Fitness

Genetic variation occurs through random mutations. Some mutations have no effect, some are harmful, and some give organisms a selective advantage—helping them survive and reproduce in a specific environment.

Table 1. Effects of Different Mutations on Bacteria

Mutation	Effect on Antibiotic Resistance	Reproductive Success
Mutation A	No resistance	Low
Mutation B	Partial resistance	Moderate
Mutation C	Full resistance	High

3. Which mutation would most likely become more common in the population over time?

(1) Mutation A
(2) Mutation B
(3) Mutation C
(4) None of the mutations will be inherited

4. Based on Table 1, explain why Mutation C would become more common in the population. How does this illustrate the concept of "differential survival and reproduction"?

____

____

____

____

Evolution Over Generations

When a population experiences natural selection, the frequency of advantageous traits increases over generations. This results in evolution—a change in allele frequency in a population over time.

Table 2. Frequency of Resistance Allele Over 10 Generations

Generation	% of Resistance Allele in Population
1	5%
3	20%
5	45%
7	70%
10	90%

5. What does Table 2 show about how resistance traits change over time?

(1) Resistance traits disappear as bacteria adapt.
(2) The resistance allele becomes less frequent due to medication.
(3) The resistance allele increases due to selective pressure from antibiotics.
(4) Antibiotic resistance is not inherited from one generation to the next.

6. Use Table 2 to explain how natural selection causes evolution in bacterial populations. Include the roles of genetic variation and selective pressure.

____

____

____

____

Human Impact on Evolution

Overuse and misuse of antibiotics (e.g., not finishing a prescription, using antibiotics for viral infections) increases the rate at which resistance develops.

Scenarios That Contribute to Antibiotic Resistance

Scenario 1: Taking antibiotics only as prescribed

Scenario 2: Stopping antibiotics early

Scenario 3: Using antibiotics for a cold (viral infection)

7. Which scenario is least likely to contribute to the evolution of antibiotic resistance?

(1) Scenario 1
(2) Scenario 2
(3) Scenario 3
(4) All scenarios equally contribute

8. Evaluate how human behavior affects the rate of evolution in bacterial populations.

____

____

____

____

9. **Claim:** "Evolution occurs as a result of natural selection acting on genetic variation." Using the evidence in this cluster, construct an argument to support this claim. Be sure to explain the cause-and-effect relationship between antibiotic use and the evolution of resistance.

____

____

____

____

Answer Explanations

1. **(3)** Antibiotic resistance is a real-world example of natural selection. Within a population of bacteria, some may carry mutations that make them resistant to an antibiotic. When antibiotics are introduced, non-resistant bacteria die off, while resistant ones survive and reproduce, passing on the resistance gene to their offspring. Over time, the population becomes dominated by resistant bacteria. This is classic survival of the fittest—the "fit" are those with traits that allow them to survive the environmental challenge (antibiotics).

2. **Sample Response:** Bacterial populations naturally have genetic variation due to mutations. Some of these mutations may provide resistance to antibiotics. When antibiotics are used, they kill the susceptible bacteria, but the resistant ones survive. These surviving bacteria reproduce rapidly, creating a new generation that also carries the resistance trait. Over many generations, this results in a population where most or all bacteria are resistant—a clear case of evolution by natural selection driven by selective pressure from the antibiotic.

3. **(3)** In Table 1, Mutation C provides the greatest level of resistance to the antibiotic and gives bacteria the highest reproductive success. Because natural selection favors traits that increase an organism's ability to survive and reproduce, Mutation C will become more common in the population over time. Mutations A and B offer less or no benefit under antibiotic stress, so bacteria with those mutations are less likely to survive and pass on their genes. This is an example of directional selection, where a specific advantageous trait becomes dominant.

4. **Sample Response:** The data show that Mutation C allows bacteria to survive exposure to antibiotics and reproduce at a higher rate than those with other mutations. This leads to differential survival and reproduction, a core principle of natural selection. Bacteria with beneficial traits are more likely to pass on those traits to the next generation, causing an increase in allele frequency over time. As a result, the population evolves to become more resistant to antibiotics.

5. **(3)** Table 2 demonstrates that the frequency of the resistance allele increases significantly over several generations, rising from 5% to 90%. This change in allele frequency over time is the definition of evolution. The data also reflect a clear cause-and-effect relationship: exposure to antibiotics increases selective pressure, favoring resistant individuals. This selective pressure drives evolutionary change in the population—a perfect illustration of evolution in action.

6. **Sample Response:** When a population of bacteria is exposed to antibiotics, the drug kills most non-resistant individuals, but any that carry a resistance gene

survive. These survivors reproduce, passing on their resistance genes. As more generations pass, the population evolves—more individuals carry the resistance allele, making the entire group harder to treat. This process is natural selection driven by human activity, and it results in an evolutionary shift in the population's genetic makeup.

7. **(1)** Correct use of antibiotics—taking the full course as prescribed—helps to ensure that all the bacteria, including partially resistant ones, are killed. If antibiotics are stopped early, the most resistant bacteria are likely to survive and reproduce, increasing resistance in future generations. Scenario 1 avoids this by eliminating the entire infection, reducing the chance for resistant bacteria to persist and multiply. This practice minimizes selective pressure, reducing the risk of resistance evolving.

8. **Sample Response:** Misusing antibiotics—such as stopping a prescription early or using antibiotics for viral infections—can increase the likelihood of resistance. In these situations, bacteria are exposed to low or inappropriate doses, which might not kill them but instead give them time to adapt and survive. These surviving bacteria may then reproduce, spreading the resistance genes. This human behavior unintentionally speeds up the rate of evolution in bacterial populations by creating favorable conditions for resistant strains.

9. **Sample Response:** Within a population of bacteria, random mutations can create variation in traits such as antibiotic resistance. When antibiotics are used, those with the resistance trait are more likely to survive. They then reproduce, passing the trait on to their offspring. Over generations, this trait becomes more common, and the population evolves. This is a classic example of natural selection acting on genetic variation, where the environment (presence of antibiotics) determines which traits are advantageous.

Topic 6: Earth's Systems

Science and Engineering Practice (SEP): Analyzing and Interpreting Data

Disciplinary Core Ideas (DCIs): ESS2.D: Weather and Climate | **LS2.C:** Ecosystem Dynamics, Functioning, and Resilience

Crosscutting Concept (CCC): Stability and Change

Phenomenon: Increasing Atmospheric Carbon Dioxide and Its Impact on Climate and Ecosystems

Base your answers to questions 1 through 9 on the information below and on your knowledge of biology.

CO_2 Levels and Climate Trends

Carbon dioxide (CO_2) is a greenhouse gas that traps heat in Earth's atmosphere. Human activities, especially the burning of fossil fuels, have increased atmospheric CO_2 levels over the past century.

Table 1. Atmospheric CO_2 Concentration and Global Temperature Anomaly (1880–2020)

Year	CO_2 (ppm)	Temp. Anomaly (°C)
1880	290	−0.2
1950	310	0.0
1980	340	+0.2
2000	370	+0.4
2020	415	+1.1

1. What relationship is best supported by Table 1?

(1) As CO_2 levels decrease, temperatures rise.

(2) As CO_2 increases, global temperature tends to rise.

(3) CO_2 levels do not affect global temperatures.

(4) Temperature change causes changes in CO_2 levels.

2. Use the data from Table 1 to explain how the increase in atmospheric CO_2 may be contributing to global climate change. Include one cause-and-effect relationship in your answer.

____

____

____

____

Carbon Movement Among Earth's Systems

Carbon cycles between the biosphere, geosphere, hydrosphere, and atmosphere through processes like photosynthesis, respiration, combustion, and dissolution.

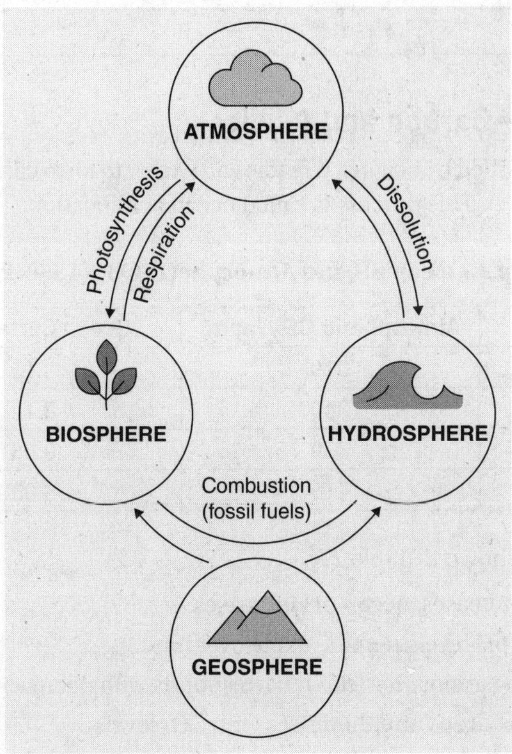

Figure 1. Simplified Carbon Cycle Model

3. Which process most directly increases atmospheric CO_2 as a result of human activity?

(1) Photosynthesis
(2) Cellular respiration
(3) Combustion of fossil fuels
(4) Uptake of CO_2 by ocean water

4. Using the model in Figure 1, describe how human actions can disrupt the natural carbon cycle. Include at least two Earth systems and one process.

____

____

____

____

Ocean Impact—Carbon and Acidity

As more CO_2 is absorbed by oceans, it reacts with water to form carbonic acid, lowering the pH of ocean water. This process is called ocean acidification.

Table 2. Ocean pH and Atmospheric CO_2 (1950–2020)

Year	Atmospheric CO_2 (ppm)	Ocean Surface pH
1950	310	8.15
1980	340	8.12
2000	370	8.08
2020	415	8.03

5. What trend is shown in Table 2?

(1) As CO_2 increases, ocean pH increases.
(2) CO_2 and pH levels remain stable over time.
(3) Increasing atmospheric CO_2 corresponds with decreasing ocean pH.
(4) Ocean pH drops and then rises with CO_2 levels.

6. Explain how ocean acidification is related to atmospheric carbon levels. How might this affect marine ecosystems?

Ecological Feedback

Ocean acidification affects shell-forming organisms like coral and plankton, which are foundational to many marine food webs. A decline in these organisms can cause disruptions throughout the ecosystem.

7. Which crosscutting concept best explains the effect of ocean acidification on marine ecosystems?
   (1) Structure and function
   (2) Energy and matter
   (3) Cause and effect
   (4) Scale, proportion, and quantity

8. Describe a feedback loop involving carbon and ocean organisms. How might a decrease in shell-forming organisms affect the carbon cycle and biodiversity?

9. **Claim:** "Human impact on Earth's systems can cause environmental changes that affect global ecosystems."

   Using information from this cluster, construct an argument supporting this claim. Include specific evidence from at least two different figures.

Answer Explanations

1. **(2)** Table 1 shows that as atmospheric CO_2 concentrations increase, the global temperature anomaly also increases. This pattern suggests a positive correlation between the two variables. CO_2 is a greenhouse gas, meaning it traps heat in Earth's atmosphere. The increase in greenhouse gases enhances the greenhouse effect, resulting in global warming. While correlation does not always imply causation, the consistent trend across decades strongly supports the climate science consensus that rising CO_2 contributes to climate change.
2. **Sample Response:** From 1880 to 2020, atmospheric CO_2 levels rose from 290 parts per million (ppm) to 415 ppm. During the same period, the global average temperature anomaly increased by over 1°C. This change suggests a direct link between rising CO_2 and climate warming. CO_2 absorbs infrared radiation and prevents heat from escaping Earth's atmosphere. This trapped heat contributes to climate change, causing disruptions such as melting ice, rising sea levels, and extreme weather.
3. **(3)** Combustion of fossil fuels—such as coal, oil, and natural gas—releases stored carbon from the geosphere (Earth's crust) into the atmosphere as CO_2. This process is the primary source of increased atmospheric carbon from human activities. Unlike natural respiration or photosynthesis, which are part of the short-term carbon cycle, fossil fuel combustion adds ancient carbon that has been stored underground for millions of years, disrupting the natural carbon balance.
4. **Sample Response:** Humans use fossil fuels for energy, releasing large amounts of CO_2 into the atmosphere through combustion. This process transfers carbon from the geosphere (where it was stored in coal, oil, or gas) to the atmosphere, increasing the concentration of greenhouse gases. Normally, the carbon cycle remains balanced, with CO_2 moving between systems (biosphere, hydrosphere, atmosphere). Human activity has tilted the balance, overwhelming Earth's natural ability to absorb carbon and leading to rising global CO_2 levels.
5. **(3)** As CO_2 levels in the atmosphere increase, more CO_2 dissolves into the ocean, where it reacts with water to form carbonic acid. This lowers the ocean's pH, making it more acidic. Table 2 clearly shows that as atmospheric CO_2 increased from 310 ppm to 415 ppm, ocean pH decreased from 8.15 to 8.03. Even small changes in pH can have serious consequences for marine life, especially calcifying organisms like coral, oysters, and plankton.

6. **Sample Response:** CO_2 dissolves into seawater and undergoes a chemical reaction to form carbonic acid (H_2CO_3). This acid breaks down into hydrogen ions, lowering the pH of the ocean and causing ocean acidification. Acidic conditions interfere with the ability of marine organisms—like coral, clams, and plankton—to form calcium carbonate shells and skeletons. When these organisms are weakened or die, the marine food web is disrupted, reducing biodiversity and undermining the overall stability of ocean ecosystems.
7. **(3)** The crosscutting concept of cause and effect is best demonstrated by this example. The cause is increasing CO_2 due to human activities, such as fossil fuel combustion and deforestation. The effect is a decrease in ocean pH, leading to harmful outcomes for marine ecosystems. Understanding cause-and-effect relationships allows scientists to make predictions and develop solutions for environmental issues like climate change and ocean health.
8. **Sample Response:** Shell-building organisms like plankton and coral play a role in sequestering carbon by converting dissolved CO_2 into calcium carbonate shells. When these organisms die, their shells can become buried in ocean sediments, effectively storing carbon long term. However, when ocean acidification interferes with shell formation, fewer organisms can sequester carbon. This keeps more CO_2 in the ocean and atmosphere, worsening climate effects and further acidifying the ocean—a dangerous feedback loop. In addition, the loss of biodiversity reduces ecosystem resilience.
9. **Sample Response:** Table 1 shows a clear trend: as CO_2 increases, global temperatures rise, linking human emissions to climate change. Table 2 shows that increased atmospheric CO_2 also causes ocean acidification, threatening marine life. These are examples of human impacts on Earth's systems, showing how one action (burning fossil fuels) can affect multiple connected systems: the atmosphere, hydrosphere, and biosphere. This highlights the importance of sustainable practices and technologies to mitigate environmental change.

Topic 7: Engineering, Technology, and Applications of Science

Science and Engineering Practice (SEP): Designing Solutions

Disciplinary Core Ideas (DCIs): ETS1.B: Developing Possible Solutions | **LS2.C:** Human Impacts on Ecosystems

Crosscutting Concept (CCC): Stability and Change

Phenomenon: Designing Solutions to Reduce Atmospheric CO_2 and Limit Climate Change

Base your answers to questions 1 through 9 on the information below and on your knowledge of biology.

Human Contribution to Rising CO_2 Levels

Burning fossil fuels for electricity, transportation, and industry releases billions of tons of CO_2 each year. This added carbon is not being balanced by natural carbon sinks (like forests or oceans), contributing to climate instability.

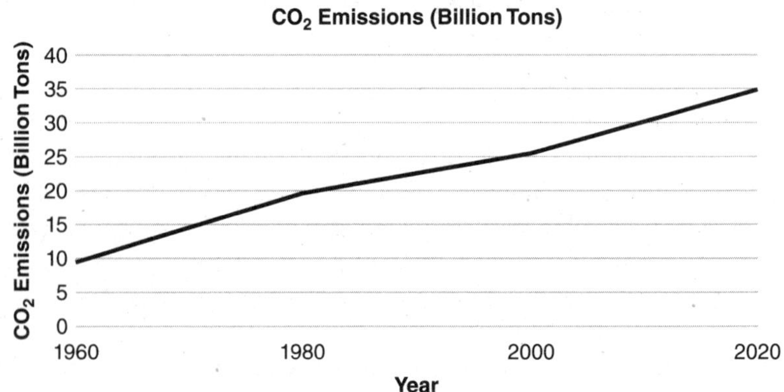

Figure 1. Global Fossil Fuel CO_2 Emissions (1960–2020)

1. Which trend is most clearly shown in Figure 1?
   (1) Global fossil fuel CO_2 emissions are decreasing steadily.
   (2) Human activity has caused CO_2 emissions to drop in recent years.
   (3) Global CO_2 emissions from fossil fuels have significantly increased since 1960.
   (4) Natural processes have eliminated excess CO_2 emissions each year.

2. Based on Figure 1, describe how human technology has contributed to changes in atmospheric CO_2. Identify a related environmental consequence.

____

____

____

____

Engineering a Solution—Carbon Capture

Engineers are exploring carbon capture and storage (CCS) as a way to reduce atmospheric CO_2. One type of CCS captures CO_2 from power plant emissions and pumps it deep underground into porous rock layers.

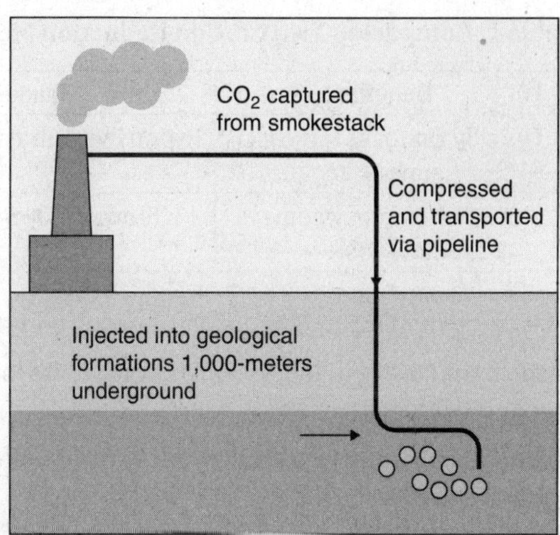

Figure 2. Diagram of a Carbon Capture and Storage System

3. Which best describes the primary purpose of carbon capture and storage?

(1) To convert CO_2 into oxygen and release it into the atmosphere

(2) To store excess carbon safely to prevent it from entering the atmosphere

(3) To increase plant photosynthesis by releasing CO_2 into forests

(4) To reduce the need for renewable energy sources

4. Explain how carbon capture and storage may help stabilize Earth's climate. Identify one potential constraint or limitation of this technology.

____

____

____

____

Evaluating Trade-Offs in Engineering Design

While CCS may reduce CO_2 emissions, it requires large infrastructure, significant energy input, and long-term monitoring. Scientists also explore natural alternatives like reforestation (planting trees) to absorb carbon over time.

Table 1. Comparing Two Carbon Reduction Methods

Strategy	Benefits	Trade-Offs
CCS	Quickly removes large CO_2 amounts	Expensive; infrastructure needed
Reforestation	Improves ecosystems; low cost	Slower CO_2 removal; land requirements

5. Which claim best compares CCS and reforestation as carbon reduction methods?
   (1) CCS is cheaper and faster than reforestation but lacks long-term results.
   (2) Reforestation is more effective than CCS at removing carbon instantly.
   (3) CCS removes more carbon quickly, but reforestation offers additional environmental benefits.
   (4) Both methods remove carbon at the same rate and cost.

6. Using Table 1, evaluate which carbon reduction method might be better in urban areas and why. Include one benefit and one trade-off.

____

____

____

____

Designing a School-Based Solution

Students at a high school propose a project to reduce their school's carbon footprint. They plan to install solar panels on the roof to replace some electricity from fossil fuels.

7. Which of the following best describes the students' project?

(1) It prevents all CO_2 from being emitted at the school.
(2) It increases fossil fuel consumption to generate more energy.
(3) It is a technological solution that reduces CO_2 emissions.
(4) It contributes to the release of methane gas into the atmosphere.

8. Explain how switching to solar energy is an example of applying science and engineering to solve a real-world problem. What is one constraint the school might face?

9. **Claim:** "Engineering solutions must balance effectiveness with environmental and economic trade-offs."

Using information from this cluster, construct an argument supporting the claim. Provide examples of both benefits and limitations of different carbon-reduction strategies.

Answer Explanations

1. **(3)** According to the data table in Figure 1, global CO_2 emissions from fossil fuels increased from 9.4 billion tons in 1960 to 34.8 billion tons in 2020. That's nearly a fourfold increase over 60 years. This dramatic rise shows the growing impact of industrialization, electricity demand, transportation, and population growth on carbon emissions. It highlights the urgency for technological solutions and policy changes to mitigate climate change.
2. **Sample Response:** Technological developments such as fossil fuel–powered vehicles, factories, and electric power plants release large quantities of carbon dioxide when burning fuels like coal, oil, and natural gas. These emissions increase the greenhouse effect, contributing to rising global temperatures (global warming). In addition, excess atmospheric CO_2 dissolves into oceans, leading to ocean acidification, which harms marine life. This illustrates how technological systems are directly connected to Earth's environmental systems—and why engineering solutions are needed to restore balance.
3. **(2)** Carbon capture and storage (CCS) is an emerging engineering solution aimed at removing CO_2 emissions at the source, such as power plants or industrial sites. The CO_2 is captured before it enters the atmosphere, compressed, and stored deep underground, usually in porous rock formations. The main goal is to reduce the net amount of greenhouse gases released by human activity. CCS is a preventative strategy to slow climate change without requiring a complete overhaul of energy infrastructure.
4. **Sample Response:** By capturing CO_2 before it reaches the atmosphere, CCS reduces greenhouse gas emissions, helping to mitigate global warming. This can be particularly effective in heavy industry and fossil-fuel-based power generation. However, there are limitations: CCS is technically complex, requires significant infrastructure investment, and must be monitored long-term to ensure that CO_2 remains safely stored. These engineering constraints—cost, safety, and scalability—must be considered when evaluating CCS as a solution.
5. **(3)** The comparison table (Table 1) shows that CCS removes large amounts of carbon quickly but has downsides such as high cost and infrastructure needs. Reforestation, on the other hand, is less expensive and provides additional benefits like habitat creation, soil stabilization, and biodiversity enhancement. However, it removes carbon more gradually. The best answer compares speed versus sustainability, showing how engineers and policymakers must weigh trade-offs when designing carbon reduction strategies.

6. **Sample Response:** In urban areas, available space for large-scale reforestation is limited, making CCS a more practical option—if funding and infrastructure allow. CCS can be integrated into existing industrial systems without requiring new land use. However, it's expensive, may face public resistance, and requires ongoing monitoring. Reforestation is cheaper and provides ecosystem benefits but isn't feasible where land is scarce. This shows how engineering design must consider context-specific constraints, such as land availability, cost, and environmental goals.

7. **(3)** Installing solar panels is a sustainable engineering solution that reduces the use of fossil fuels by capturing renewable solar energy. This reduces the school's carbon footprint, aligning with modern goals for green technology and sustainability. Unlike fossil fuels, solar energy does not emit CO_2 during use. The students' proposal is a perfect example of how engineering can solve real-world problems while considering environmental impacts.

8. **Sample Response:** Solar panels use renewable energy from sunlight to generate electricity, reducing reliance on fossil fuels and lowering carbon emissions. This is a practical way to apply science and engineering for environmental benefit. However, solar projects often face constraints, such as the high upfront cost of installation, need for sufficient roof space, and weather dependency. These are typical challenges engineers must address when scaling up solutions for real-world applications.

9. **Sample Response:** CCS is highly effective at capturing CO_2 quickly, but it's expensive, energy-intensive, and requires secure geological sites for storage. Reforestation takes longer but is cost-effective, improves biodiversity, and restores natural systems. Solar energy avoids emissions entirely but requires initial investment and access to sunlight. Each solution has strengths and trade-offs. Engineers must consider not only effectiveness, but also economic feasibility, environmental benefits, and long-term stability when designing and implementing these systems.

Chapter 7

Regents Practice Exams

Regents Practice Exam June 2025

Directions: **For each multiple-choice question, record in the space provided the number of the choice that best completes the statement or answers the question. For questions requiring a written response, write your answer in the spaces provided.**

Base your answers to questions 1 through 5 on the information below and on your knowledge of biology.

Carbon! Where Does It Come From? Where Does It Go?

On Earth, carbon compounds are found in the oceans, atmosphere, and living organisms, as well as stored in rocks and sediments. Earth and its atmosphere can be considered a closed system. The amount of carbon in different locations within Earth's system is always changing.

Sea otters help maintain the carbon balance in their ecosystem. Otters eat sea urchins. Eating sea urchins is important, as sea urchins are herbivores that can destroy a kelp forest. Kelp are large autotrophic algae that grow much faster than most plants. When kelp die, they sink into the deep ocean. The low oxygen conditions of the sea floor cause decomposition to be slow or incomplete.

Scientists calculated the carbon pool (how much carbon kelp stores) with and without sea otters, as shown in the model below.

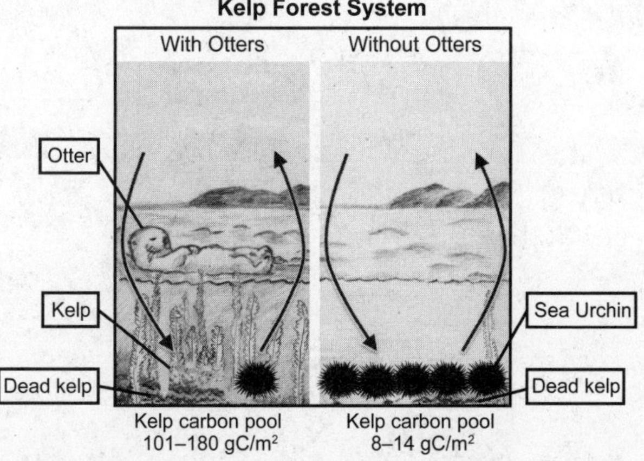

1. Which claim about the kelp carbon pool is best supported by evidence from the information and the model above?

(1) The carbon storage is higher with the otters present because the sea otters eat the sea urchins.

(2) The carbon storage is higher with the sea urchins present because they control the kelp population.

(3) The carbon storage is lower with the otters present because the sea otters eat the kelp.

(4) The carbon storage is lower with the sea urchins present because they carry out autotrophic nutrition.

1 ______

2. Which statement uses the model to describe how kelp contributes to a reduction of carbon entering the atmosphere?

(1) Kelp produces carbon as it grows within the hydrosphere.

(2) Some of the carbon stored in the dead kelp is trapped in the geosphere of the sea floor.

(3) Kelp produces carbon as it sinks into the hydrosphere.

(4) Some of the carbon stored in kelp is added to the geosphere through cell respiration.

2 ______

3. Which model identifies the process that converts light energy into chemical energy inside the kelp?

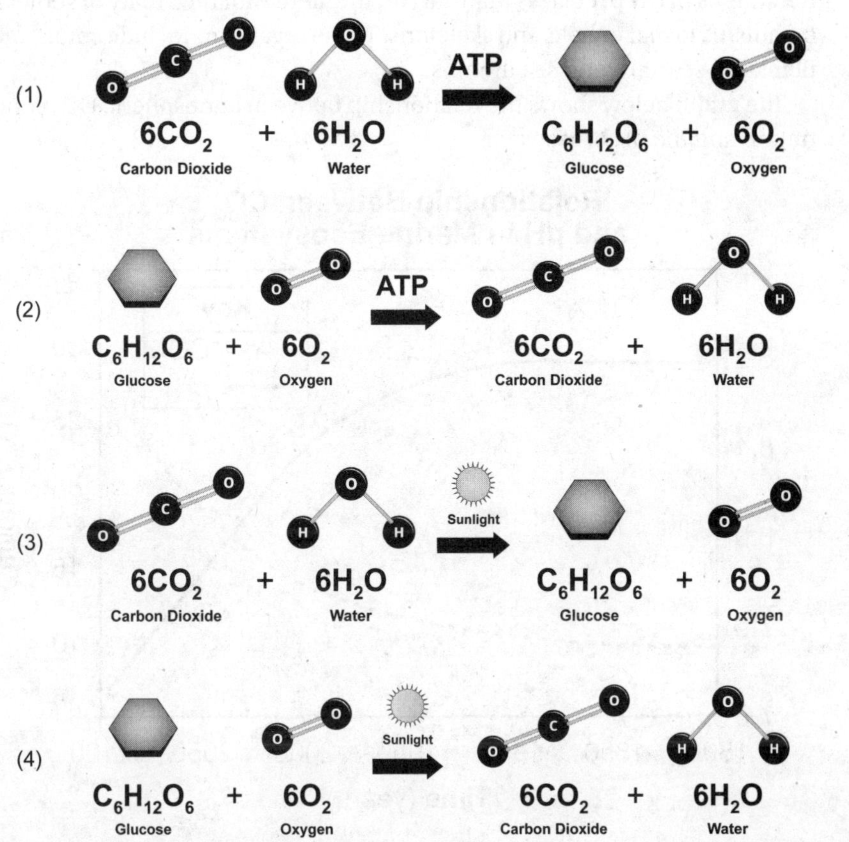

3 ______

Increased atmospheric carbon dioxide has been linked to changes within marine ecosystems. When CO_2 combines with water, it produces carbonic acid, lowering the water's pH. A pH of less than 7.8 can interfere with the ability of some marine organisms to make shells and skeletons. These organisms include corals, mussels, plankton, seastars, and sea urchins.

The graph below shows the relationship between atmospheric CO_2 concentration and the pH of seawater.

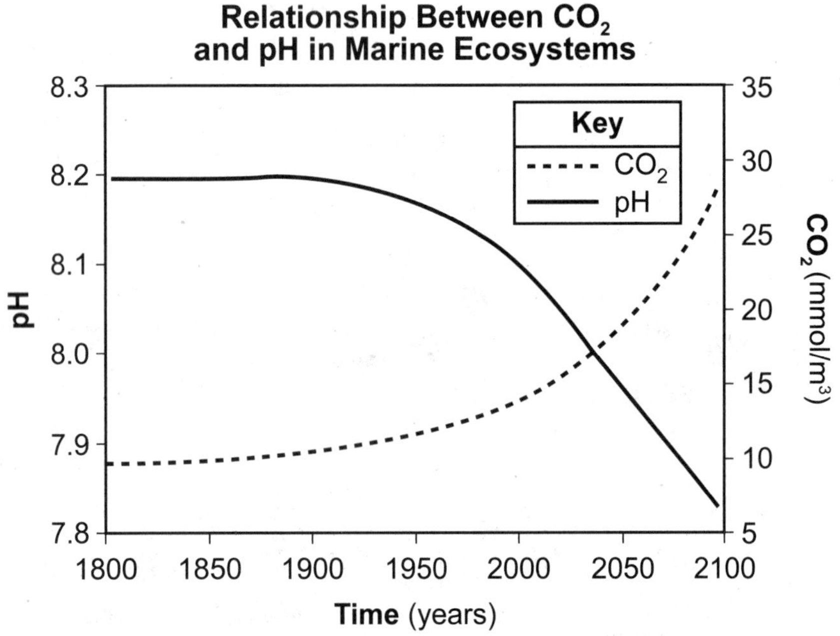

4. If the trend in atmospheric CO_2 levels continues, sea urchin populations may be impacted. Describe evidence from the graph that supports this claim. [1]

____

____

____

____

____

____

The following diagram shows some information about the cycling of carbon.

Model of Ocean Acidification

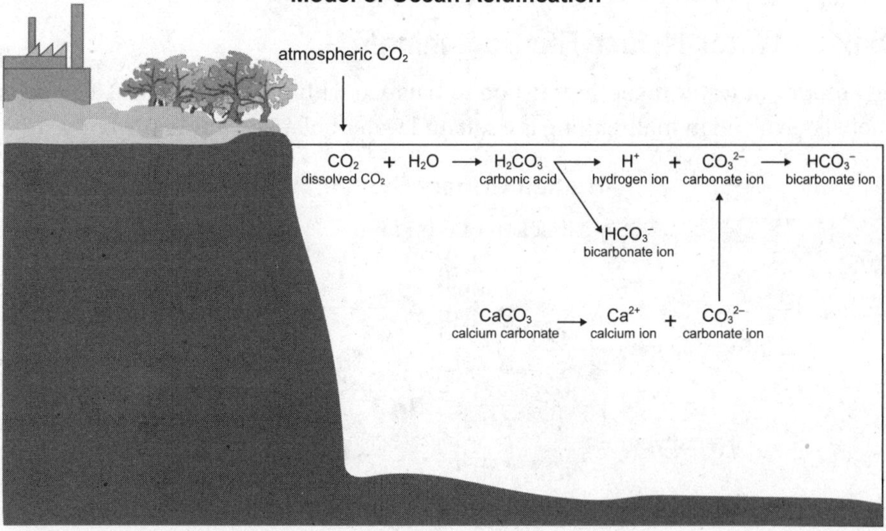

atmospheric CO_2

CO_2 (dissolved CO_2) + H_2O → H_2CO_3 (carbonic acid) → H^+ (hydrogen ion) + CO_3^{2-} (carbonate ion) → HCO_3^- (bicarbonate ion)

H_2CO_3 (carbonic acid) → HCO_3^- (bicarbonate ion)

$CaCO_3$ (calcium carbonate) → Ca^{2+} (calcium ion) + CO_3^{2-} (carbonate ion)

(Model not to scale)

The model below shows the equation for how sea urchins make their shells.

$$Ca^{2+} + CO_3^{2-} \rightarrow CaCO_3$$

Calcium ions and carbonate ions produce calcium carbonate.

5. As ocean acidification increases, the amount of available carbonate ions decreases. Use the model and information provided to describe how the cycling of carbon between the biosphere *and* at least one other sphere is affected as environmental conditions change. [1]

____

____

____

____

____

____

Base your answers to questions 6 through 10 on the information below and on your knowledge of biology.

Drinking Water Is Just The Beginning!

The amount of water taken in must be in balance with the amount lost. The urinary system is involved in maintaining the salt and water balances within the body.

Human Urinary System

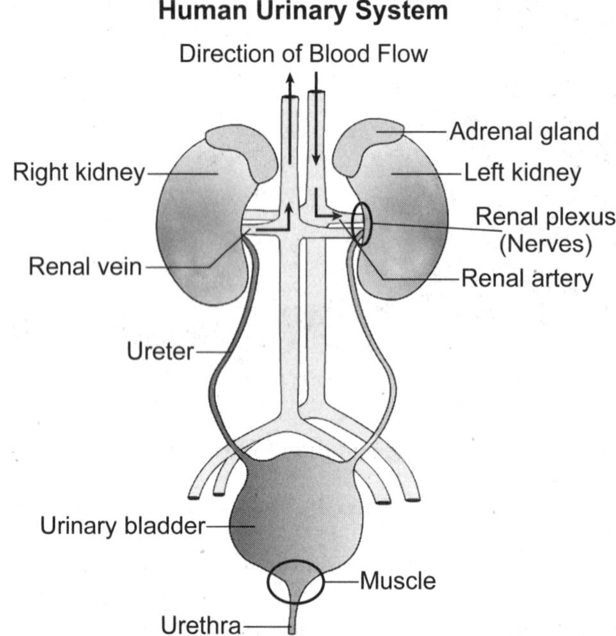

6. Which statement describes how the organization of the urinary system and *one* other system interact to maintain homeostasis in the human body?
   (1) The adrenal gland, part of the endocrine system, delivers nutrients to the cells of the urinary system to remove carbon dioxide from the blood.
   (2) The internal urethral sphincter muscle, part of the muscular system, contracts to signal the cells of the urinary system to regulate blood sugar.
   (3) The brain, part of the nervous system, sends messages to the renal plexus (nerves) to signal the cells of the urinary system to deliver oxygen to the blood.
   (4) The arteries, part of the circulatory system, deliver unfiltered blood to the cells of the urinary system to remove wastes.

6 ______

Each kidney is composed of about a million waste-filtering structures known as nephrons. Water is reabsorbed through some parts of the nephron, such as the tubule. The model below shows the structure of a nephron in two organisms.

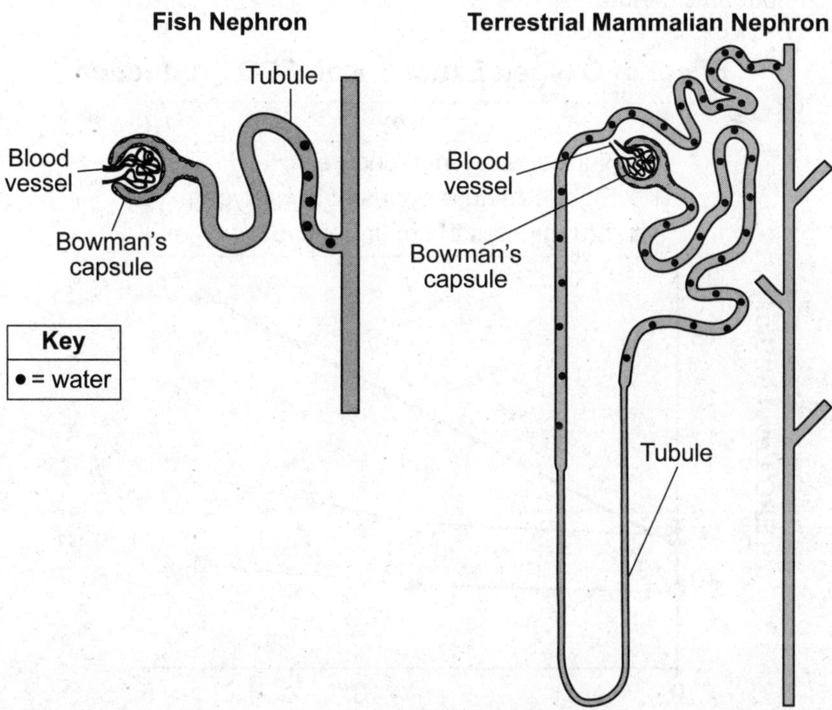

7. Based on evidence in the model, which statement explains that natural selection led to the evolution of the nephron's structure and function in terrestrial mammals?

(1) The development of Bowman's capsule in animal kidneys was more advantageous to mammals than fish.

(2) Nephrons with longer tubules were selected for in organisms that lived on land to conserve the water they drank.

(3) The longer nephron tubule was selected for in animals that lived in oceans, rivers, and lakes to filter the excess water absorbed.

(4) The number of nephrons is more important in the evolution of mammals than it is to the evolution of fish.

7 ______

Another function of the kidney is to aid in regulating the number of red blood cells. The kidneys produce a protein known as erythropoietin (EPO) that stimulates the increased production of red blood cells.

The graph below shows the results of a study. Participants were exposed to different conditions prior to time zero, then EPO levels were measured over a 4.5-hour time period.

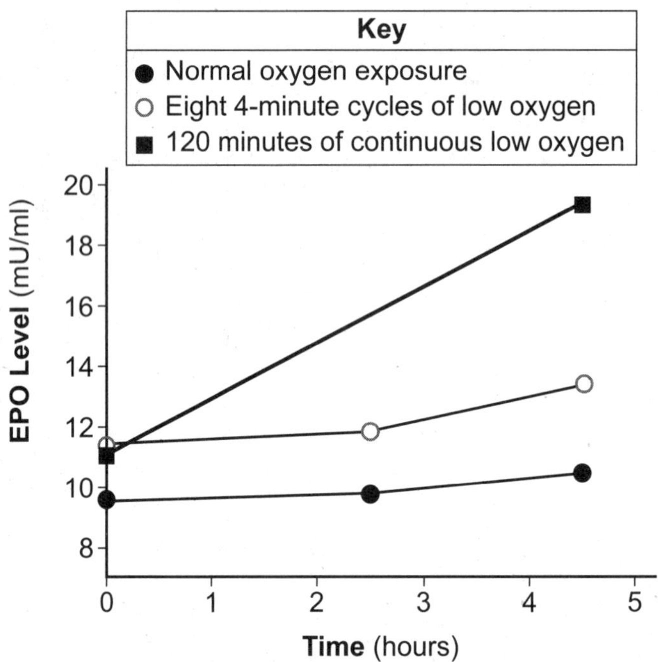

8. Using the information provided, describe the evidence to support the claim that exposure to low oxygen levels results in a feedback mechanism that allows the body to maintain homeostasis. [1]

____

____

____

____

In addition to their other functions, red blood cells (RBCs) have the capacity to carry water. The surface of the cell contains structures called aquaporins, which transport water across cell membranes. Due to their unique cell structure, RBCs are able to expand up to 74% or shrink up to 40% compared to the original cell size.

9. Which statement provides the best explanation for how red blood cells contribute to a feedback mechanism that maintains homeostasis?

(1) The aquaporins in the red blood cells facilitate the exchange of water in environments of different concentrations and help the kidneys more efficiently regulate salt and water balance.

(2) The aquaporins in the red blood cells prevent the exchange of water in environments of different concentrations and help the kidneys more efficiently regulate salt and water balance.

(3) The red blood cells change their shape to move through the kidney but have minimal effect on salt and water balance in the kidneys.

(4) The red blood cells change their shape to move through the kidney but do not regulate any feedback mechanisms.

9 ______

RBCs typically have a biconcave disk shape. A scientific study examined the effect of changing the RBC shape and membrane flexibility on its ability to transfer oxygen. For this study, when the RBC membrane was less flexible, the oxygen transfer capacity decreased by 18%. When the RBC was more flexible, the oxygen carrying capacity increased by 21%. When RBCs take on water, their membrane initially becomes more flexible.

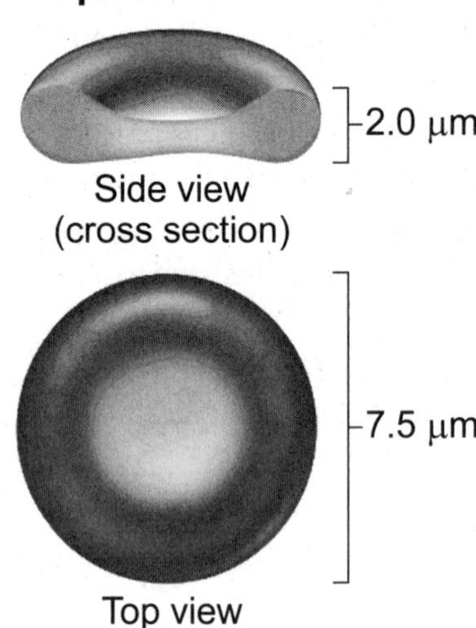

10. What evidence would support the claim that consuming water after exercise helps an athlete's body maintain homeostasis?

	Number of RBCs Taking on Water	RBC Flexibility	Rate of Oxygen Transfer
(1)	increases	decreases	decreases
(2)	increases	increases	increases
(3)	decreases	increases	increases
(4)	decreases	decreases	decreases

10 ______

Base your answers to questions 11 through 16 on the information below and on your knowledge of biology.

Heads or Tails?

In the mid-1990s, people across several states were finding large numbers of frogs and other amphibians with extra limbs. Possible explanations regarding the cause of these abnormalities ranged from UV radiation, chemical contaminants in the water, parasites, or even airborne substances.

11. Which question could be asked in order to determine if the abnormalities seen in the frog legs were caused by an inherited mutation?

(1) Do the offspring with an abnormality live in the same environment as the parents?

(2) Were the parents exposed to the same environmental factors as some of their offspring?

(3) Is a mutation that causes abnormal legs present in the DNA of the sex cells of the parents?

(4) Do the cells within the legs of the parents contain DNA with a mutation that causes abnormal limbs?

11 ______

After further research, scientists discovered that these deformities in frogs were not caused by genetic mutations. The actual cause was a parasitic flatworm called *Ribeiroia. Ribeiroia* completes a complex life cycle by inhabiting several hosts. This life cycle is summarized in the diagram below.

***Ribeiroia* Life Cycle**

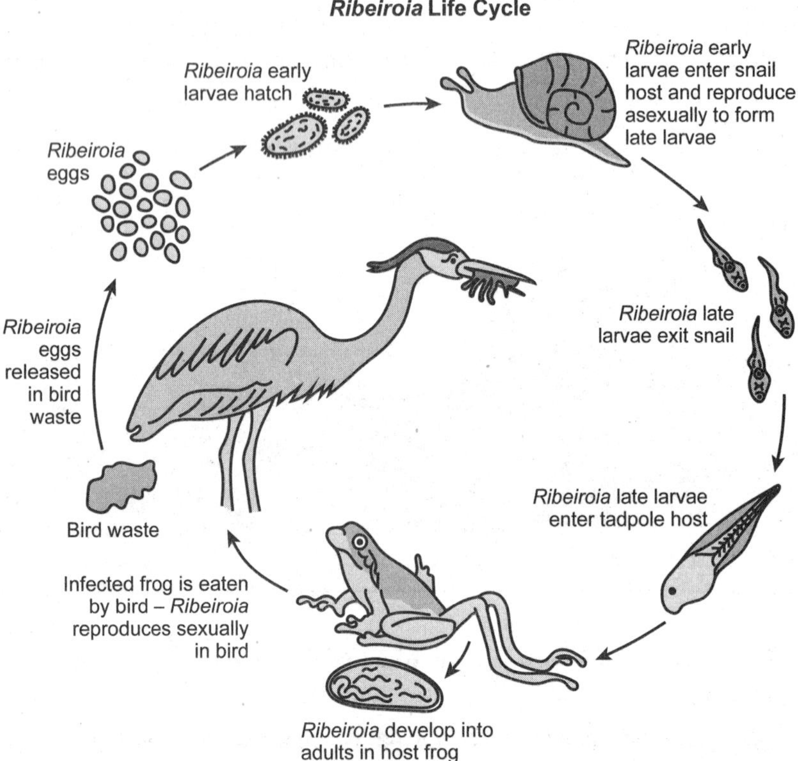

12. A student claims that the *Ribeiroia* parasites that cause the most severe limb abnormalities in frogs have a greater chance of survival and reproduction than those that do not. Which explanation best supports this claim?

    (1) The frogs with the most severe limb deformities will be more likely to be caught by birds, allowing the adult *Ribeiroia* to be more likely to survive and reproduce.

    (2) The adult *Ribeiroia* will have a greater chance of remaining in the frog to complete all phases of its life cycle, allowing it a better chance to survive and reproduce.

    (3) The *Ribeiroia* will have a greater chance of reproducing sexually because it stays in the snail, releasing larvae with this trait back into the water.

    (4) The *Ribeiroia* larva will have a greater chance of reproducing asexually and completing its life cycle in the bird.

12 ______

Hox genes are an important group of regulatory genes that help determine the body plan and head-to-tail orientation of animals in their early stages of development. High concentrations of retinoic acid have been found to influence the activity of Hox genes.

The diagram below shows how proteins produced by the activated Hox genes attach to DNA sequences that act as molecular switches to turn large numbers of different genes on.

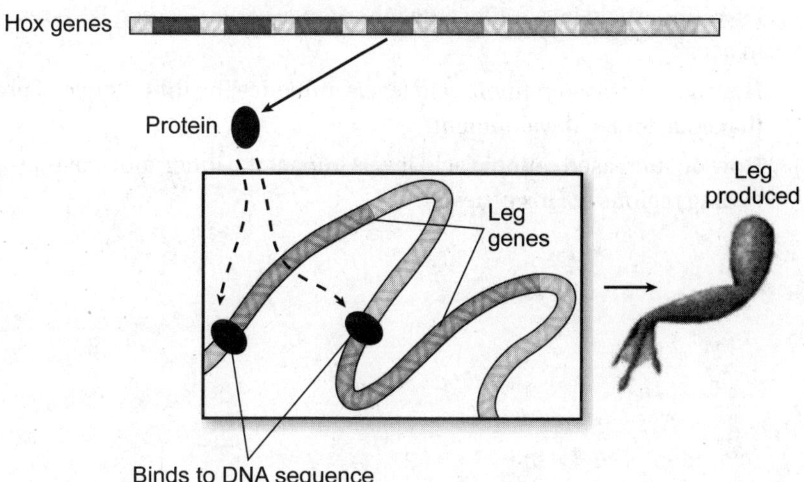

13. Using the information above, which statement might best explain why the extra limbs grew in parasite-infected frogs?

(1) The Hox genes in the limb caused an increase in the retinoic acid levels, which produced proteins that signaled the leg genes to turn off.

(2) The parasites increased the level of retinoic acid in the tadpole limb, causing Hox genes to transcribe more of the proteins that activated limb formation genes.

(3) The proteins produced by the developing limb signaled the Hox genes to turn on, which increased the retinoic acid levels, causing more legs to grow.

(4) The higher retinoic acid levels caused by the parasites turned off the Hox genes in the tadpole limb, signaling the leg formation genes to activate.

13 ______

Scientists have found that when the *Ribeiroia* parasite enters the frog tadpole, it burrows into the limb bud, which develops into the frog's leg. The levels of a chemical called retinoic acid increase rapidly in the limb bud of the tadpole due to the parasitic infection.

14. Which question could be asked about the effect of retinoic acid levels on inheritance of the observed changes in the frogs?

(1) Do retinoic acid levels affect the inheritance of Hox genes that code for proteins important for leg development?

(2) Do retinoic acid levels affect the inheritance of non-coding DNA resulting in a leg?

(3) How do decreased retinoic acid levels influence the inheritance of proteins that code for leg development?

(4) How do increased retinoic acid levels impact the inheritance of non-coding regions of Hox genes?

14 ______

Hox genes are also found in arthropods. The diagrams below show some information about the Hox genes and body segmentation in a fruit fly. The body plans for other arthropod species are also shown. The differing shades of gray indicate the Hox genes responsible for each body segment's development.

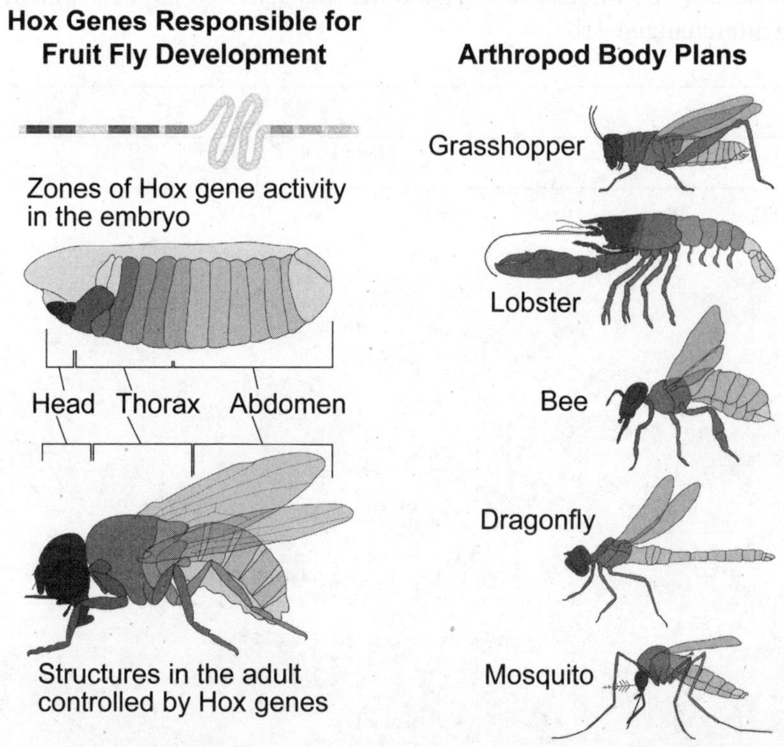

15. Describe genetic *and* physical evidence that would support the claim that all of these arthropods share a common ancestor. [1]

____

____

____

Hox genes are also present in mammals and other vertebrates to produce specific body parts in the correct orientation. Specific Hox genes from a mouse and fruit fly can be interchanged.

16. Construct an explanation for why typical functioning eyes form in a mouse and a fly when specific Hox genes that activate eye development are interchanged. [1]

____

____

____

Base your answers to questions 17 through 21 on the information below and on your knowledge of biology.

Organisms of the Tibetan Plateau

The yak is an herbivore that lives in the high altitude of the Tibetan Plateau, located in the Himalayan Mountains. They inhabit the Tibetan Plateau at elevations between 3,000 and 5,000 meters. Yaks have a large heart and lungs, as well as a specialized hemoglobin in the cells of their blood that enables them to extract more oxygen from the air.

Concentration of Atmospheric Oxygen Available at Different Altitudes

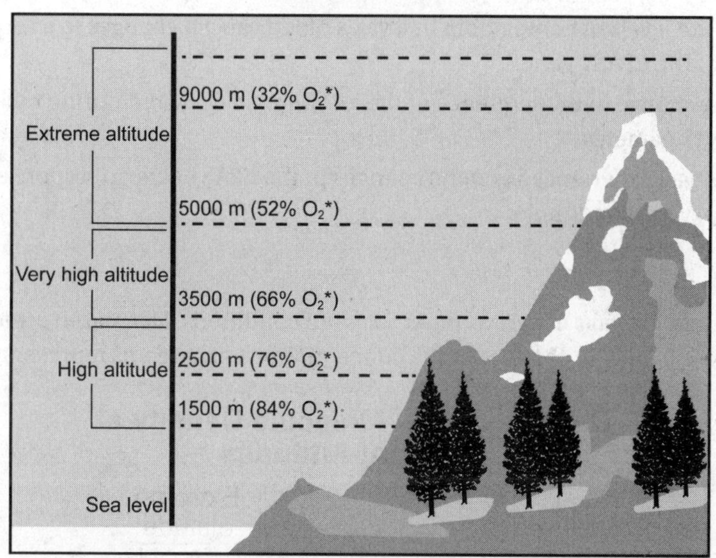

*Percentage of oxygen (O_2) available at this altitude compared to sea level

17. Using evidence provided, construct an explanation that describes how natural selection has led to the development of an adaptation within yak populations that enable them to survive in their environment. [1]

____

____

____

____

A gene called EPAS1 is involved in allowing animals to respond to a low oxygen environment. Scientists studied this gene in yaks. They found that yaks having a certain allele of the gene had a greater amount of hemoglobin, which transports oxygen through the body. This allele has a small change in nucleic acid sequence from other alleles of the EPAS1 gene found in yak populations.

18. What is the most likely origin of the change that resulted in this allele?

(1) A change in the sequence of the EPAS1 gene occurred during the mitosis of yak blood cells.

(2) The levels of hemoglobin in a yak's blood caused changes to a sequence of the EPAS1 gene.

(3) A change in the sequence of the EPAS1 gene occurred during meiosis of yak gametes.

(4) A yak experienced genetic changes in the EPAS1 gene in response to low oxygen conditions.

18 ______

Plants that yaks eat grow under stressful conditions. Decreased atmospheric pressure results in changes in the concentrations of gases, as represented below.

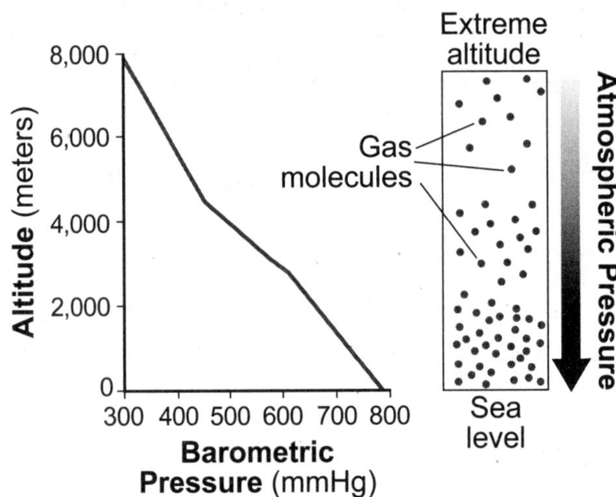

19. Which statement best explains why the carrying capacity for producers in extreme-altitude ecosystems is *less* than in sea-level ecosystems?

(1) There is more oxygen available to use for cellular respiration at high altitudes.
(2) There is less carbon dioxide available to use for photosynthesis at extreme altitudes.
(3) There is increased water vapor available at extreme altitudes, which limits the process of photosynthesis.
(4) There is decreased pressure at high altitudes, resulting in faster cellular respiration.

19 ______

The pika is another herbivorous mammal species that inhabits the Tibetan Plateau. Pika move quickly and spend much of their time foraging for food and keeping a lookout for predators. They are small (5–9 inches long), live in systems of underground tunnels that they dig and maintain, and lack a large heart and large lungs.

The photograph and graph below show some information about the Tibetan Plateau.

Relationship Between Altitude and Temperature

Air temperature (°C): 16, 14, 12, 10, 8, 6, 4, 2, 0, −2, −4

Altitude (m): 2000, 2500, 3000, 3500, 4000, 4500, 5000

20. Construct an explanation based on evidence that natural selection leads to a behavioral adaptation that pika possess that would help them to survive in the Tibetan ecosystem. [1]

____

____

____

____

Other organisms that are part of the Tibetan Plateau ecosystem include carnivores such as wolves, eagles, and snow leopards.

The model below shows some approximate biomass at each trophic level of the Tibetan Plateau ecosystem.

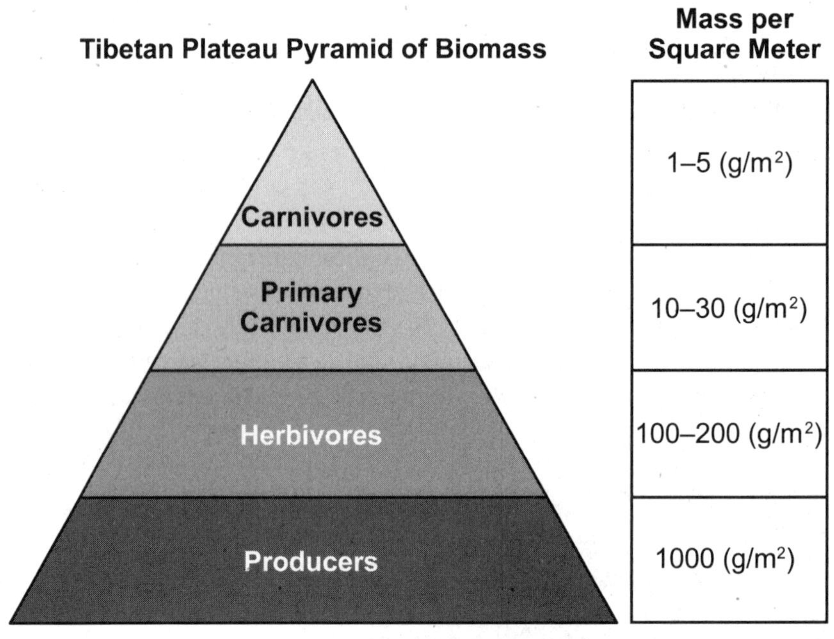

21. Use the evidence provided to make a claim about how energy flow among organisms in the Tibetan Plateau ecosystem affects biomass. [1]

____

____

____

Base your answers to questions 22 through 27 on the information below and on your knowledge of biology.

Salt Marsh Shoreline

Salt marshes are unique ecosystems located along an ocean shoreline, between the ocean itself and dry, upland ecosystems. They are important areas that filter water, protect coastlines, and provide essential habitat. Salt marshes can be affected by tides and weather events. Depending on a variety of factors, salt marshes can contain varied amounts of vegetation, which can influence the biodiversity and function of the salt marsh. The model below shows some information about a typical salt marsh.

Anatomy of a Salt Marsh

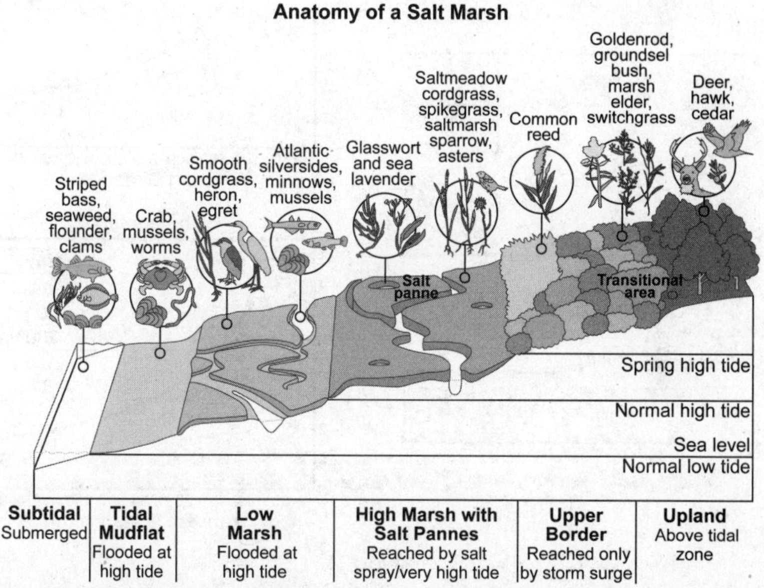

22. Which claim best describes the complex interactions that would have the greatest immediate effect on populations of smooth cordgrass in the low salt marsh?

(1) The smooth cordgrass population would be most affected by an influx of saltmarsh sparrows that relocate after severe storms.

(2) The smooth cordgrass population would be most affected by rising sea levels due to global warming that affects the tide level.

(3) The smooth cordgrass would be most affected by a temporary increase in salinity due to severe storms.

(4) The smooth cordgrass would be most affected by erosion due to severe storms and resulting floods.

22 ______

Erosion can affect shorelines, including salt marshes. Mathematical models often report shoreline erosion using transect distance. Transect distance measures the same sand dunes along the same line between two specific points. Due to the influence of tides on coastal zones, elevation also affects erosion. The models below show some information about factors that affect shorelines.

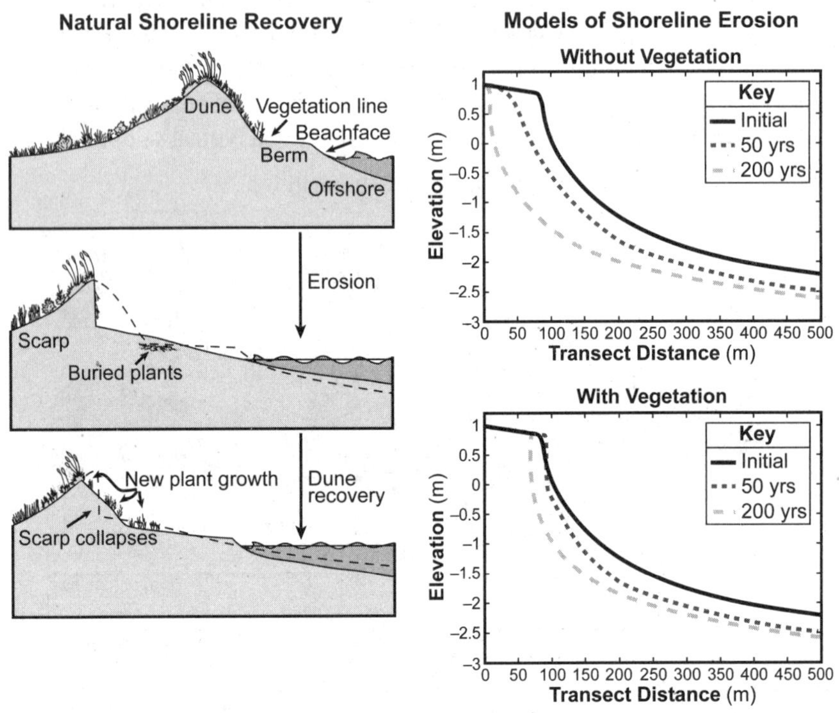

23. Using the information provided, which statement best describes how natural shoreline erosion affects the habitat's carrying capacity at different scales?

(1) Habitat carrying capacity rapidly decreases in shorelines with vegetation that are affected by erosion.

(2) Habitat carrying capacity rapidly increases in shorelines with vegetation that are affected by erosion.

(3) Habitat carrying capacity rapidly decreases in shorelines without vegetation that are affected by erosion.

(4) Habitat carrying capacity rapidly increases in shorelines without vegetation that are affected by erosion.

23 ______

Global climate change may impact salt marshes and other coastal ecosystems. The graph below shows some data collected by NASA using satellites.

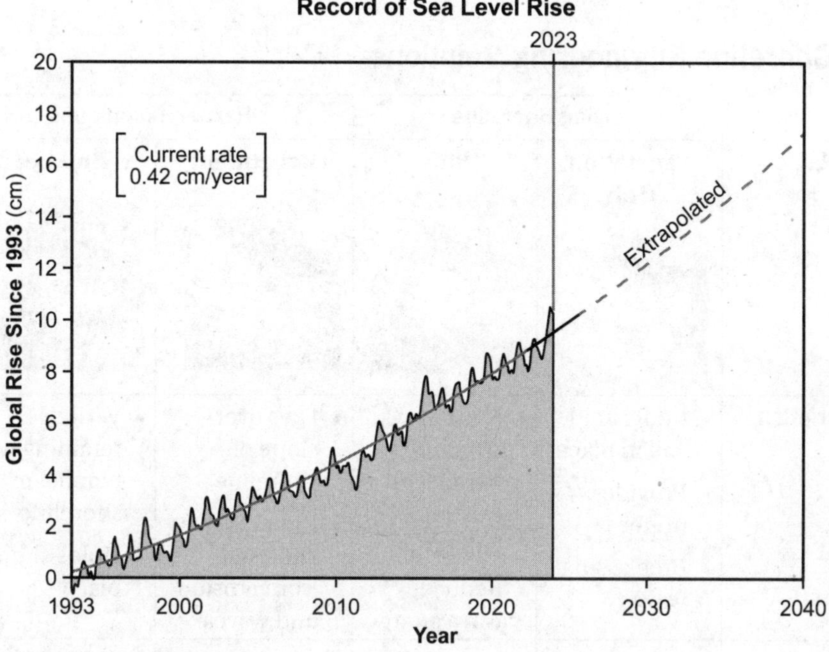

24. Using the evidence provided, evaluate the claim that rising sea levels will impact the numbers and types of organisms that interact within a low marsh. [1]

____

____

____

Strategies that mimic the natural surroundings are being developed to reduce erosion and restore shoreline ecosystems. The chart below shows some information about the characteristics of various options for shoreline restoration.

Shoreline Engineering Solutions

	Living Shoreline		Harder Techniques	
Name	**Vegetation Only**	**Sills**	**Revetment**	**Bulkhead**
Description	• Roots hold soil in place • Provides buffer • Breaks small waves	• Natural structures parallel to existing habitat • Reduce wave energy	• Lays over slope of shoreline • Protects shoreline from erosion and waves	• Vertical retaining wall parallel to shoreline • Holds shore in place
Optimal Conditions	Low energy wave environments	Low to moderate wave environments	Sites with preexisting hardened shorelines	High energy wave environments
Material Options	Native plants	Stone and living reef (oysters, mussels)	Stone, rubble, concrete blocks or slabs, sand/concrete-filled bags	Steel, timber, concrete, carbon fiber
Advantages	• Provides habitat • Slows inland water transfer and stores water • Maintains aquatic/terrestrial connection	• Provides habitat • Slows upland water transfer • Prevents wetland loss • Natural barriers to waves	• Reduces wave action • Low maintenance • May fragment habitat	• Moderates wave action • Reduces tide fluctuations • Prevents natural marsh migration

	Living Shoreline		Harder Techniques	
Name	**Vegetation Only**	**Sills**	**Revetment**	**Bulkhead**
Disadvantages	• No high water protection • Vegetation may not grow	• No high water protection • Vegetation may not grow	• No major flood or high water protection • Loss of aquatic/terrestrial connection	• No major flood protection • Loss of aquatic/terrestrial connection • Decreases fisheries' habitat and diversity
Cost Initial Construction	$	$$	$$$$	$$$
Operations & Maintenance Construction	$	$	$$	$$

25. Which type of restoration project would promote the conditions necessary for the development of a stable ecosystem that would support complex interactions between the organisms living there?

(1) The formation of a salt marsh area would allow only one species of water plant to grow in the damaged area.

(2) Build bulkheads so fish and water plants would be held back from shorelines.

(3) Planting vegetation along damaged shorelines would provide shelter for the protection of the shoreline organisms.

(4) The building of a hard structure, such as a revetment, would increase the energy from the waves striking the shoreline.

25 ______

26. Which claim best describes the complex interactions in a natural living shoreline that are affected by changing conditions, such as a severe storm?

(1) Erosion that occurs during a storm can cause changes to the slope of the land, which exposes high marsh plants to decreased levels of salt, reducing biodiversity.

(2) Oyster reefs can reduce erosion caused by storm waves, which allows salt marshes to expand upland, increasing habitat availability for other organisms.

(3) Bulkheads prevent erosion and increase biodiversity by preventing marsh migration and maintaining the berm during storms.

(4) Severe storms have no impact on the salinity of a salt marsh, which causes cordgrass to die, resulting in shorelines that have less vegetation to prevent erosion.

26 ______

Water levels across the Great Lakes are primarily the result of natural, uncontrolled water supplies into the basin. In June 2019, Lake Ontario experienced record high water levels resulting from heavy rains and storms. Oswego, NY is a city located on the shoreline of Lake Ontario.

Significant damages and other impacts were experienced across the system. Concerns about shoreline loss ranged from loss of revenue from recreational activities in lakeside towns, such as boating, fishing, swimming, and fine dining, to property loss for homeowners and businesses. Shoreline communities are seeking reliable ways to reduce property damages and maintain the beach town culture and natural recreational opportunities.

The photograph shows one shoreline in Oswego, NY after the 2019 high water event.

27. Identify the best possible option for shoreline restoration in Oswego, NY from the Shoreline Engineering Solutions chart based on the associated cost, reliability, and aesthetics. Evaluate the social and environmental impacts of this solution using those criteria and their trade-offs. [1]

____

____

____

____

____

____

____

Base your answers to questions 28 through 32 on the information and diagrams below and on your knowledge of biology.

Limb Evolution

Tetrapods include all animals with backbones that have four limbs that end with digits (fingers and toes). Some tetrapods, such as whales and snakes, do not have four obvious limbs but are included because they have a four-limbed ancestor.

The front limbs of tetrapods are believed to have evolved from the pectoral fins of an ancestral bony fish.

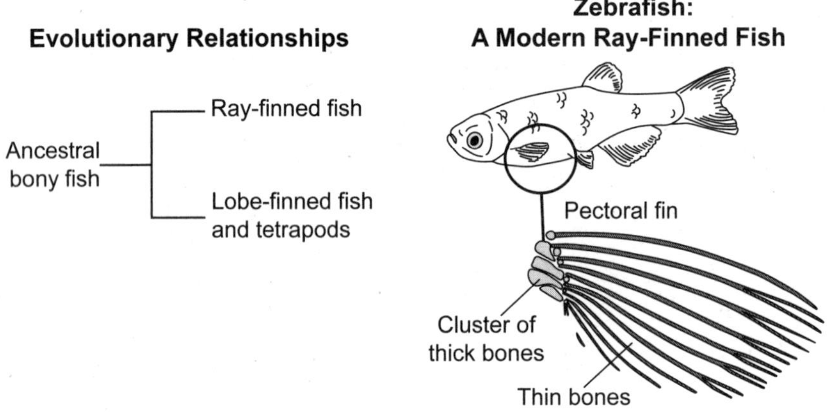

28. What evidence could be used to support the claim that there are patterns in how front limbs evolved in descendants of ancestral bony fish?

(1) Documented changes in the habitat of ancestral ray-finned fish that required them to change the number of genes used to produce pectoral fins

(2) Similarities in the base sequence of genes that control the development of pectoral fins in the zebrafish and front limbs in tetrapods

(3) A comparison of the total number of fins on a zebrafish and the total number of limbs present on a tetrapod that is alive today

(4) Information regarding how the front limb is used within the environment of the modern-day tetrapod

28 ______

The diagram summarizes some of the current structural and fossil information regarding front limb evolution in some living animal species and some extinct water-dwelling animal species.

Evolution of Tetrapod Front Limbs

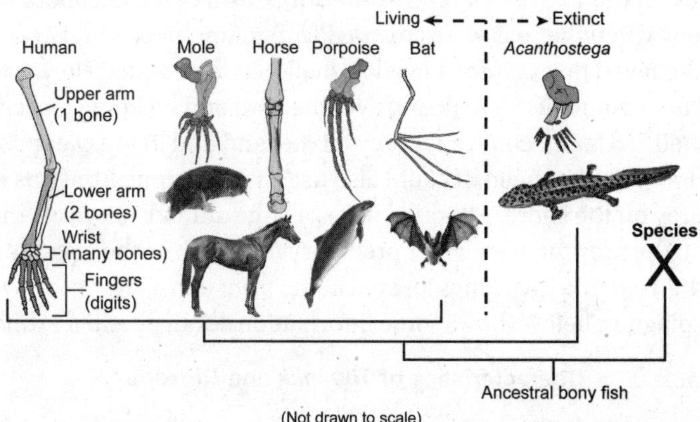

(Not drawn to scale)

29. Which statement below identifies the evolutionary relationships presented in the diagram?

(1) The forelimbs of *Acanthostega* and the living species have a bone structure best suited to life on land; therefore, the forelimbs of all species evolved from an extinct land-dwelling ancestor.

(2) The extinct, water-dwelling *Acanthostega* and porpoise share the most similar habitat; therefore, they share the most recent, extinct common ancestor.

(3) Each of the living species has different forelimb bone structures because they developed different structures in order to evolve within their specific habitats.

(4) The forelimbs of the extinct, water-dwelling species and the living species have a similar arrangement of bones, which provides evidence of common ancestry.

29 _____

30. Construct an explanation, based on evidence, that the evolution of limb development can be the result of environmental factors. [1]

____

____

____

____

Scientists are searching for transitional fossils to provide evidence that land-living tetrapods evolved from bony fish. This missing species is represented by Species X on the Evolution of Tetrapod Front Limbs diagram. In 2004, fossilized remains of a possible contender were discovered in Canada. It was named *Tiktaalik*. *Tiktaalik* was a large, fish-like organism that lived about 385 million years ago, when the seas were crowded with many species of fish.

It is believed that *Tiktaalik* lived in shallow, warm water. During this time period the first plants were taking over the land and creeping insects and spiders flourished. *Tiktaalik* could spot prey on the land and in the water using its eyes, located on top of its head. It could also use its very strong front fins to chase and catch prey on the shore. Although it was a large animal, it is likely that *Tiktaalik* was also the prey for even larger predatory fish, such as the gigantic *Rhizodus*, which had two massive fangs located at the front of its jaw.

The diagram below shows some information about possible extinct organisms.

Characteristics of *Tiktaalik* and *Rhizodus*

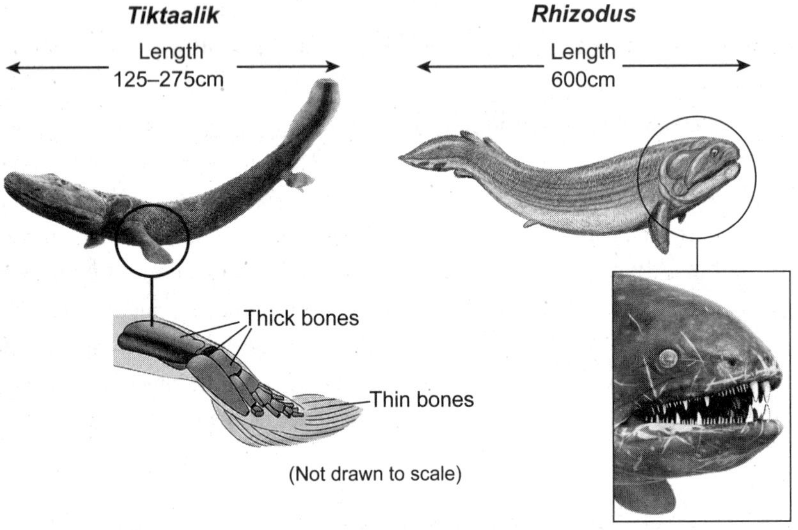

31. Use patterns in the front limb bone structure to support the researcher's claim that *Tiktaalik* represents an ancestral form between ray-finned fish and the early tetrapod, *Acanthostega*. [1]

____

____

____

32. What evidence supports the explanation that animals with traits well-suited to life on land evolved because of the environmental factors present 385 million years ago?

(1) The large body of *Rhizodus* allowed it to move quickly through the shallow water.

(2) The location of *Tiktaalik's* eyes allowed it to see prey in both the land and water.

(3) *Rhizodus* had large fangs that allowed it to function as a predator of *Tiktaalik* both on land and in the water.

(4) *Tiktaalik* was able to access new food sources and evade predation by *Rhizodus* because the bone structure of its fins enabled it to walk on land.

32 ______

Base your answers to questions 33 through 37 on the information below and on your knowledge of biology.

Does It Matter?

The carbon used by plants is moved between living organisms, the minerals in the soil, the hydrosphere, and the atmosphere through processes in the carbon cycle.

Matter Conversions

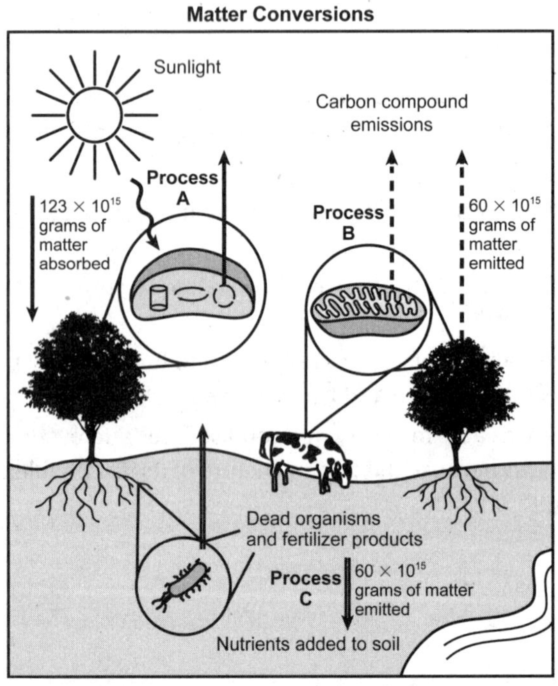

33. Using information from the model, which statement correctly identifies how the movement of matter in this ecosystem provides energy for different organisms?

(1) The plant takes in carbon compounds from the atmosphere, which get converted into sugar and then used by the cow during Process *B* to produce usable energy.

(2) The plant takes in oxygen from the soil, which gets converted into nutrients during Process *B* to produce usable energy.

(3) The cow performs Process *C*, which releases sugars into the atmosphere and then is used by plants during Process *A* to produce usable energy.

(4) The cow's wastes are broken down through Process *A*, which releases sugars into the soil and then is used by plants during Process *C* to produce usable energy.

33 ______

34. Using evidence from the model, construct an explanation for the role of Process *C* in the cycling of matter between living organisms in this ecosystem. [1]

____

____

____

Plants rearrange matter to produce other needed compounds. The model below shows some of the compounds that plants synthesize. Boxes *X*, *Y*, and *Z* represent element(s) used to make these compounds.

How Plants Rearrange Matter

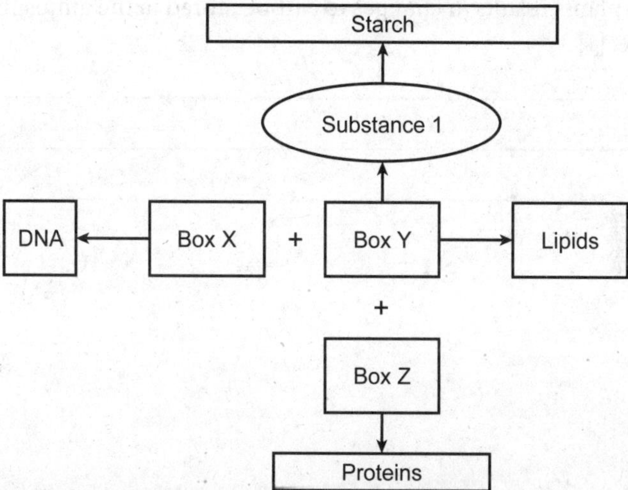

35. Which explanation best supports the claim that elements from Substance 1 in the model combine with different elements to form other carbon-based molecules?

(1) The elements in Box *Y* are broken down into nitrogen and phosphorus and then combined to form lipids.

(2) Substance 1 molecules can be joined together to make starch.

(3) The elements in Box *Y* are combined with nitrogen to make the substances that are used to form proteins.

(4) Substance 1 molecules can be joined together to make DNA.

35 ______

36. Based on the information in all the models provided, which claim can be made about why Substance 1 is essential for plant metabolism?

(1) Process *B* combines Substance 1 with other elements to form lipids to be utilized by the plant.

(2) Process *A* rearranges elements of carbon, hydrogen, and oxygen to form Substance 1 to be utilized by the plant.

(3) Processes *A* and *C* combine nitrogen and phosphorus with Substance 1 to form proteins utilized by the plant.

(4) Processes *B* and *C* rearrange nitrogen and Substance 1 to form DNA and starches utilized by the plant.

36 ______

37. Construct an explanation using quantitative evidence for how the cycling of matter in plants results in changes to carbon stored in the atmosphere and biosphere. [1]

____

____

____

Base your answers to questions 38 through 42 on the information below and on your knowledge of biology.

Keystone Species: Black-Tailed Prairie Dogs

The black-tailed prairie dog is a keystone species because it maintains the complex web of relationships in North America's central grassland ecosystems. They feed primarily on plants that are high in moisture and nutrients. As they eat the plants, they drop leaf clippings, which add nutrients to the soil. They construct burrows, which when abandoned can provide homes to rattlesnakes, burrowing owls, and insects. Prairie dogs are the primary prey for many organisms, including the black-footed ferret, one of the rarest and most endangered animals in North America.

The prairie dog population of North America's central grassland is on a steady decline. The most significant threats facing prairie dogs are the conversion of rangeland to cropland, urban development, hunting, and use of poison because they are considered a pest to local farmers and ranchers.

The graph below shows some data collected in a study of 17 prairie dog colonies in Nebraska.

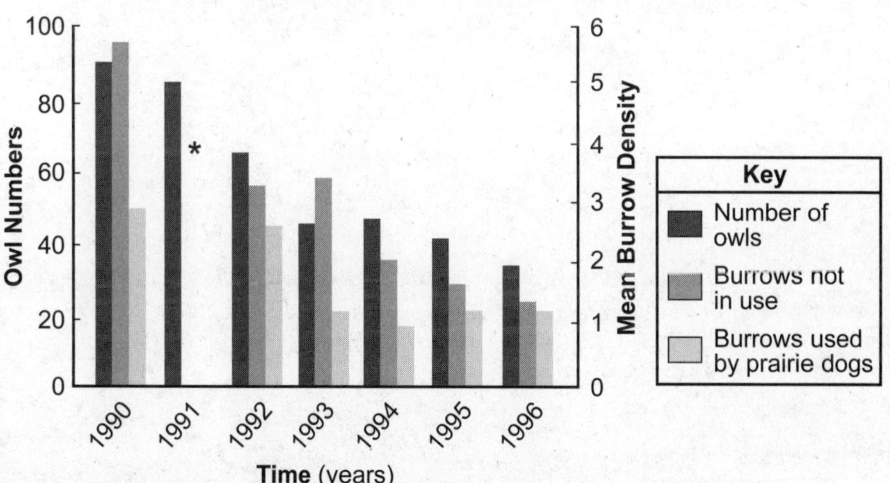

*Complete data not available for 1991.

38. How did the number of prairie dog burrows affect the carrying capacity of burrowing owls in the area?

(1) As the number of prairie dog burrows decreased, the number of burrowing owls the area could support increased.

(2) As the number of burrowing owls increased, the number of prairie dogs the area could support decreased.

(3) As the number of prairie dog burrows decreased, the number of burrowing owls the area could support decreased.

(4) As the number of burrowing owls increased, the number of total burrows the area could support decreased.

38 ______

39. Evaluate the claim that a significant decrease in the prairie dog population would have widespread effects by identifying a specific interaction between ecosystem components. [1]

____

____

____

A disease known as the Sylvatic Plague is caused by a bacterium that is carried by fleas on rats. The disease targets small mammals, including the prairie dog. This disease entered the western United States as a consequence of the shipping industry and has been spreading eastward. The graph below shows changes in the prairie dog populations observed in two different states during the 1990s.

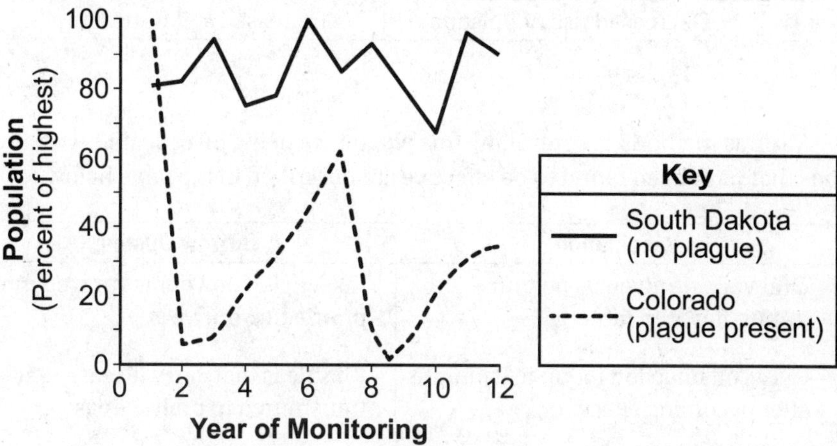

40. A claim was made that some of the prairie dogs in Colorado had an advantageous heritable trait that protected them from the plague. Which statement would provide evidence to support this claim?

(1) The prairie dog population increased between years 1 and 2 and years 7 and 8, but then declined because the prairie dogs with the protective variation soon died.

(2) The prairie dog population declined between years 1 and 2 and years 7 and 8, but was able to recover because the prairie dogs with the protective variations survived to reproduce.

(3) The prairie dogs were protected from the plague because the percentage of prairie dogs surviving the plague was always above 60.

(4) Both populations of prairie dogs recovered from plague infections before year 12.

40 ______

41. The South Dakota prairie dog population shows fluctuation within a range. Which row of the table identifies how different factors affect the carrying capacity?

Row	Factor keeping population from dropping substantially	Factor keeping population from increasing substantially
(1)	Urban development	Number of abandoned burrows
(2)	Depletion of soil nutrients	Reduced rangeland
(3)	Preservation of grassland	Predation by ferrets
(4)	Decreased use of poisons	Increased soil nutrients

41 ______

Various methods of controlling this plague are being investigated. Two methods that have been found to be effective are described in the table below.

Vaccination	Burrow Dusting
- Oral vaccine given as peanut butter-flavored tablet	- Insecticide powder is sprayed into prairie dog burrows
- Fights off infection for up to 9 months after becoming effective	- Kills fleas that carry disease that is transmitted to prairie dogs
- Tablets must be eaten by the prairie dogs within 7 days of being dropped	- Can reduce fleas for up to 2 years beginning immediately after spraying

The location of the study is near residential areas and open ranges that are used to graze cattle and serve as habitats for wild animals. Researchers have been asked to make recommendations regarding which strategy would be the best to use to protect the prairie dog populations from the plague without negatively impacting the nearby areas.

42. Describe the treatment, vaccination or burrow dusting, that will best protect the prairie dogs from the plague while considering the criteria and constraints of cost, safety, *or* reliability. Use specific information from the table to justify your choice of cost, safety, *or* reliability. [1]

____

____

____

____

Base your answers to questions 43 through 48 on the information below and on your knowledge of biology.

Nature or Nurture?

During one winter in the 20th century, the Netherlands experienced a severe famine (a shortage of food). Some of the women that were in the early stages of pregnancy during the famine gave birth to children that surprisingly had average or even above-average birth weights, considering the poor nutrition of the mothers.

43. Which question would help determine the role of DNA in passing genetic information that influenced birth weight from the mothers during the famine to their children?

(1) Did the genes that play a role in determining birth weight come from both parents?
(2) Did genes made of amino acids come from the DNA of only one parent?
(3) Did genes composed of protein come from the DNA of both parents?
(4) Did the stomach cells of the mother have genes that play a role in birth weight?

43 ______

Scientists determined that the children whose mothers were in early pregnancy during the famine (children of the famine) had more obesity and chronic health problems in adulthood compared to their siblings who did not have this exposure. The children of the famine had changes in expression of some of their genes. One of these genes, known as IGF2 (insulin-like growth factor 2), codes for a hormone.

The codon chart below can be used to determine the amino acids that are coded for by a DNA sequence.

Codons in mRNA

First Base	Second Base								Third Base
	U		C		A		G		
U	UUU	Phenylalanine	UCU	Serine	UAU	Tyrosine	UGU	Cysteine	U
	UUC	Phenylalanine	UCC	Serine	UAC	Tyrosine	UGC	Cysteine	C
	UUA	Leucine	UCA	Serine	UAA	Stop	UGA	Stop	A
	UUG	Leucine	UCG	Serine	UAG	Stop	UGG	Tryptophan	G
C	CUU	Leucine	CCU	Proline	CAU	Histidine	CGU	Arginine	U
	CUC	Leucine	CCC	Proline	CAC	Histidine	CGC	Arginine	C
	CUA	Leucine	CCA	Proline	CAA	Glutamine	CGA	Arginine	A
	CUG	Leucine	CCG	Proline	CAG	Glutamine	CGG	Arginine	G
A	AUU	Isoleucine	ACU	Threonine	AAU	Asparagine	AGU	Serine	U
	AUC	Isoleucine	ACC	Threonine	AAC	Asparagine	AGC	Serine	C
	AUA	Isoleucine	ACA	Threonine	AAA	Lysine	AGA	Arginine	A
	AUG	Methionine or start	ACG	Threonine	AAG	Lysine	AGG	Arginine	G
G	GUU	Valine	GCU	Alanine	GAU	Aspartic Acid	GGU	Glycine	U
	GUC	Valine	GCC	Alanine	GAC	Aspartic Acid	GGC	Glycine	C
	GUA	Valine	GCA	Alanine	GAA	Glutamic Acid	GGA	Glycine	A
	GUG	Valine	GCG	Alanine	GAG	Glutamic Acid	GGG	Glycine	G

A portion of the IGF2 DNA sequence is included in the table below.

DNA	CTC	CAC	GCT
mRNA	GAG	GUG	CGA
Amino Acid	Glutamic Acid	Valine	Arginine

44. A student claimed that changing CTC to CTG in the DNA would result in production of a different protein. Which explanation supports this claim?

(1) When GAG turns to GAC, aspartic acid is included in the protein instead of glutamic acid.

(2) When GAG turns to GAC, there is no change to the protein produced.

(3) When GAG turns to GAC, the protein would include valine in place of glutamic acid.

(4) When GAG turns to GAC, all of the amino acids that make up the protein would be different.

44 ______

In a cell, DNA interacts with different molecules. It also interacts with methyl groups in a process known as methylation.

Methylated vs Unmethylated DNA

Methylated DNA

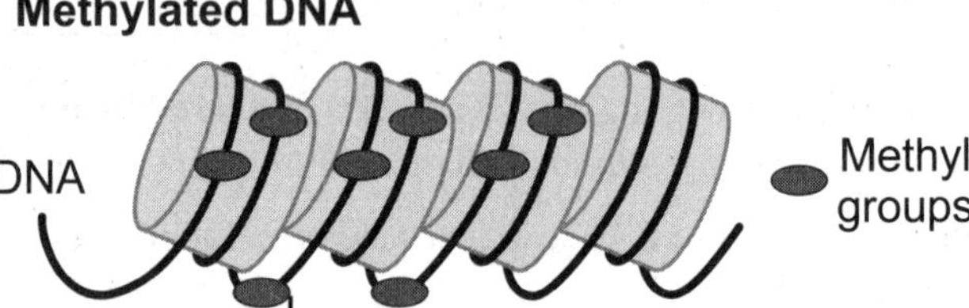

Unmethylated

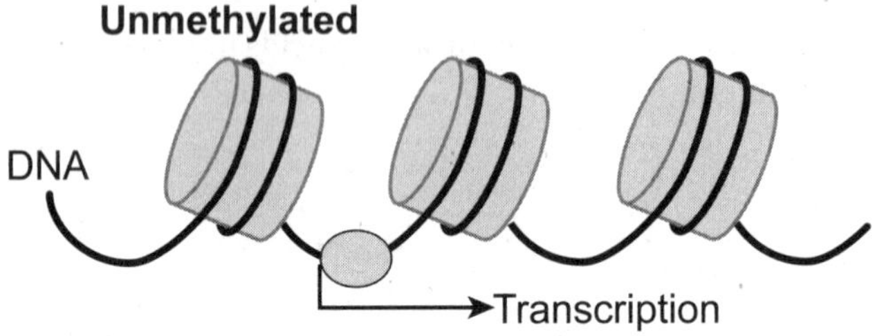

The IGF2 gene codes for a hormone that promotes fetal growth. The children of the famine had less methylation of the IGF2 gene than is typical in other children.

45. Construct an explanation based on evidence for how the structure of unmethylated DNA impacts the function of IGF2, resulting in greater birth weight. [1]

____

____

____

Researchers tracked the health of the children of the famine and their offspring for many years. The research findings showed similar results in their offspring.

46. Which claim is best supported by evidence that their offspring had similar health problems despite a lack of exposure to famine?

(1) Methylation levels of DNA are inherited for only one generation.
(2) Methylation levels of DNA can be inherited over many generations.
(3) The DNA base sequence is the only factor affecting gene expression.
(4) The DNA base sequence is protected from mutation due to methylation.

46 ______

Besides famine during pregnancy, other environmental factors such as smoking nicotine can cause demethylation (removal of methyl groups) of portions of DNA responsible for cell division. The result of this demethylation is shown in the model below.

Effects of Demethylation on Lung Cells

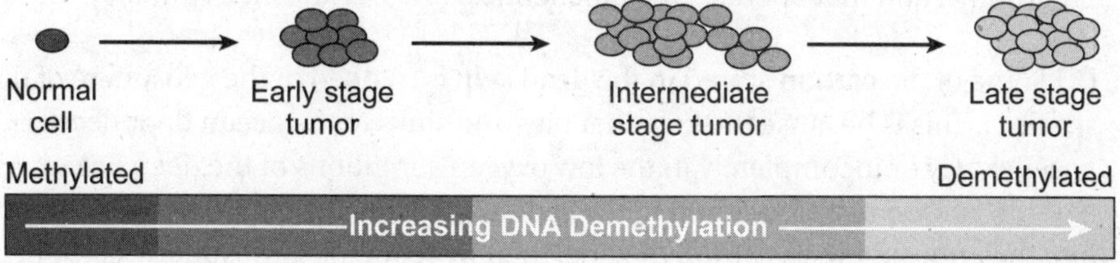

47. Using evidence from the model, describe how the disruption of the flow of information affects lung cells with demethylated DNA. [1]

____

____

____

48. Which statement identifies a solution that researchers could use to decrease tumor growth and progression?

(1) Use radiation to remove methyl groups from the tumor genes.
(2) Use gene therapy to modify the DNA to accelerate cell division in tumor cells.
(3) Use medication to add methyl groups to the genes that cause an increase in cell division.
(4) Use medication that increases the rate of mitosis in all body cells.

48 ______

Answer Explanations June 2025

1. **(1)** The cell membrane is the organelle in a single-celled organism that functions in a similar way to the respiratory system of an elephant. Within the elephant's lung, oxygen is absorbed, and carbon dioxide is released through cell membranes of the alveoli. In a like manner, oxygen and carbon dioxide are exchanged through the cell membrane of the single-celled organism.

Wrong Choices Explained

(2) The nucleus is an organelle in a single-celled organism that contains the genetic material of the cell and that controls the cell's biochemical processes.

(3) The vacuole is an organelle in a single-celled organism that collects, stores, and excretes some of the cell's biochemical wastes.

(4) The chloroplast is an organelle in a single-celled organism that contains chlorophyll and that operates the biochemical process known as photosynthesis.

2. **(2)** Some of the carbon stored in the dead kelp is trapped in the geosphere of the sea floor. This is because the kelp that dies and sinks to the ocean floor decomposes slowly or incompletely in the low oxygen conditions of the deep ocean. As a result, carbon remains stored in the sediment instead of being released back into the atmosphere as carbon dioxide, thereby reducing atmospheric carbon.

Wrong Choices Explained

(1) Kelp does not produce carbon as it grows. It removes carbon dioxide from the atmosphere through the process of photosynthesis and stores it in its tissues.

(3) Kelp does not produce carbon as it sinks. It stores carbon absorbed earlier during photosynthesis.

(4) Cell respiration does not add carbon to the geosphere. It releases carbon dioxide into the atmosphere.

3. **(3)** The third model correctly shows the process of photosynthesis, which converts light energy into chemical energy inside the kelp. The correct balanced equation is $6CO_2 + 6H_2O + \text{light energy} \rightarrow C_6H_{12}O_6 + 6O_2$, which shows carbon dioxide and water being converted into glucose and oxygen using sunlight. This represents how kelp, as an autotroph, stores energy chemically.

Wrong Choices Explained

(1) This model is incorrect because it either lacks essential components like sunlight or misrepresents the chemical equation.

(2) This model likely shows products of respiration (ATP and glucose), not photosynthesis.

(4) Although it contains some correct parts, this version is neither a complete nor an accurate model of photosynthesis.

4. **Sample Response:** Sea urchin populations may be impacted by the continuing trend of increasing atmospheric CO_2 levels because the data in the graph show a clear inverse relationship between atmospheric CO_2 and seawater pH. As CO_2 increases, seawater pH drops below 7.8. This acidity impairs the ability of organisms like sea urchins to build calcium carbonate shells and skeletons.

5. **Sample Response:** As ocean acidification increases, carbonic acid forms more readily, releasing hydrogen ions that bind with carbonate ions. Fewer carbonate ions remain for organisms like sea urchins to build calcium carbonate structures. This reduces the cycling of carbon into the biosphere (via shells) and limits the storage of carbon in ocean sediments (geosphere).

6. **(4)** The arteries, part of the circulatory system, deliver unfiltered blood to the kidneys so that waste products can be removed from the blood. This coordination between the circulatory and urinary systems maintains homeostasis by removing metabolic waste and regulating water and salt balance.

Wrong Choices Explained

(1) The endocrine system influences kidney function, but adrenal glands do not deliver nutrients.

(2) The urethral sphincter controls urination, not blood sugar regulation.

(3) The nervous system may regulate kidney function, but the kidneys do not oxygenate blood.

7. **(2)** Nephrons with longer tubules in terrestrial mammals evolved through natural selection because they help conserve water—a critical advantage on land. Mammals that retained more water through these extended tubules had higher survival and reproductive success in dry environments.

Wrong Choices Explained

(1) Both nephron types have Bowman's capsule; this was not the selected feature.

(3) Longer tubules are not advantageous in aquatic environments; short tubules help eliminate excess water.

(4) This choice incorrectly focuses on nephron quantity over adaptive structural features.

8. **Sample Response:** The graph shows that after exposure to low oxygen, EPO levels increase over time. This supports the claim that the body uses a feedback mechanism—sensing low oxygen and responding by producing more erythropoietin (EPO), which stimulates red blood cell production. This restores oxygen delivery and helps maintain homeostasis.

9. **(1)** Red blood cells (RBCs) use aquaporins to move water in and out depending on the surrounding environment. This water exchange helps regulate cell volume and maintain salt-water balance, which supports the kidneys in preserving internal stability (homeostasis).

Wrong Choices Explained

(2) Aquaporins enable—not prevent—water exchange.

(3) RBC shape helps in transport, but RBCs also play an active role in fluid balance

(4) RBCs are involved in homeostatic mechanisms, particularly water transport.

10. **(2)** After exercise, drinking water helps restore hydration, which allows red blood cells (RBCs) to absorb water. This rehydration increases the fluidity and flexibility of their membranes, improving their ability to navigate through narrow blood vessels and deliver oxygen efficiently. This improves oxygen transfer rates by up to 21%, as shown in the study. This adaptation helps the athlete recover faster and maintain homeostasis by efficiently delivering oxygen to cells.

Wrong Choices Explained

(1) Increased water would not decrease flexibility or oxygen transport.

(3) Water intake does not reduce RBC numbers or function.

(4) Water consumption increases—not decreases—oxygen transport capacity.

11. **(3)** To determine if the limb deformities in frogs were inherited, scientists must check if a mutation exists in the DNA of the sex cells. Only mutations in gametes can be passed from parent to offspring genetically.

Wrong Choices Explained

(1) Living in the same environment does not determine inheritance.

(2) Exposure to an environmental factor (such as a chemical or parasite) may cause deformities, but this does not confirm a genetic mutation in the sex cells. Environmental exposure is not evidence of a heritable genetic change, so it does not prove that the deformities can be inherited.

(4) Somatic cells (like leg cells) do not pass mutations to offspring.

12. **(1)** This choice is supported by the *Ribeiroia* life cycle. Frogs with severe limb deformities are more vulnerable to predation by birds. Since the *Ribeiroia* parasite completes sexual reproduction in birds, deformities that increase predation help the parasite survive and reproduce more effectively.

Wrong Choices Explained

(2) *Ribeiroia* completes sexual reproduction in birds, not frogs.

(3) Asexual reproduction occurs in snails, not birds.

(4) Larvae reproduce asexually in snails, not birds, and this does not relate to severity of deformities.

13. **(2)** Hox genes regulate the body plan of animals. In this case, the parasite likely raised levels of retinoic acid, which increased the activity of Hox genes. The activated Hox genes produced proteins that turned on limb development genes, leading to the growth of extra legs in the infected frogs.

Wrong Choices Explained

(1) Retinoic acid affects Hox gene expression, not the other way around.

(3) Proteins from the limb do not activate Hox genes; Hox genes control limb development.

(4) Turning off Hox genes would not cause extra limb growth.

14. **(1)** This question investigates whether retinoic acid influences the inheritance of Hox genes. Since Hox genes direct limb formation, understanding if retinoic acid affects gene inheritance is essential to determine if the trait could be passed to offspring.

Wrong Choices Explained

(2) Non-coding DNA is not directly responsible for traits like leg development.

(3) Decreased levels of retinoic acid were not the concern in this study.

(4) The focus is on gene inheritance, not on non-coding regions.

15. **Sample Response:** Genetic evidence such as similarities in Hox gene sequences and functions across arthropods supports a common ancestor. Physical evidence, like similar segmented body structures controlled by Hox genes, further strengthens this claim. For example, all arthropods shown in the diagram have head, thorax, and abdomen sections, indicating shared ancestry.

16. **Sample Response:** The fact that a Hox gene from a mouse can activate eye development in a fruit fly shows that the genetic code for eye development is highly conserved across species. This means the Hox genes responsible for turning on eye-related genes work similarly in both animals, despite species differences.

17. **Sample Response:** Natural selection favored yaks with traits that helped them survive in low-oxygen environments. These yaks had larger hearts and lungs and a specialized form of hemoglobin that allowed more efficient oxygen uptake. These adaptive traits became more common in the population over generations as yaks with these traits reproduced more successfully.

18. **(3)** The most likely origin of the EPAS1 allele variation in yaks is a mutation during meiosis in gametes. This type of mutation can be passed to offspring, making it a heritable trait that can spread through natural selection if it provides a survival advantage in low-oxygen environments.

Wrong Choices Explained

(1) Mitosis affects body cells and cannot pass traits to offspring.

(2) Environmental conditions like hemoglobin levels do not directly cause genetic changes.

(4) Genetic changes are not a direct response to environmental conditions in individuals.

19. **(2)** At higher altitudes, there is less atmospheric pressure and thus fewer carbon dioxide molecules available for photosynthesis. Since carbon dioxide is essential for making glucose, reduced CO_2 limits the number of producers an ecosystem can support.

Wrong Choices Explained

(1) Oxygen is used in respiration, not in photosynthesis.

(3) Water vapor tends to decrease at higher altitudes.

(4) Cellular respiration is not faster due to decreased pressure.

20. **Sample Response:** One behavioral adaptation of the pika is that it digs and lives in underground tunnels. This behavior helps them survive in the Tibetan Plateau by providing shelter from extremely cold temperatures at high altitudes and protection from predators. Natural selection favored individuals that dug tunnels because they were more likely to survive and reproduce in this harsh environment.

21. **Sample Response:** The pyramid shows that producers have the greatest biomass and carnivores have the least. This happens because energy is lost at each step of the food chain, so less energy is available to support organisms at higher trophic levels. As a result, the biomass decreases from producers to herbivores to carnivores in the Tibetan Plateau ecosystem.

22. **(4)** Erosion from storms physically removes sediment and plants, such as smooth cordgrass, from the low marsh. This has an immediate impact on the smooth cordgrass population by reducing available habitat and exposing roots.

Wrong Choices Explained

(1) Bird movement has less direct effect on cordgrass.

(2) Rising sea level is gradual, not immediate.

(3) Salinity changes may affect growth but are not the most immediate threat.

23. **(3)** Without vegetation, shoreline elevation decreases more rapidly over time due to erosion. This leads to habitat loss and a reduction in the area's ability to support diverse life forms, thus lowering carrying capacity.

Wrong Choices Explained

(1) Vegetation helps slow erosion and preserve habitats.

(2) Erosion does not increase carrying capacity.

(4) Shorelines without vegetation are less stable.

24. **Sample Response:** The graph shows that sea level has been rising steadily since 1993 and is projected to continue rising in the future. As sea levels rise, low marshes may become flooded more often, which can reduce the number of plant species able to survive there. This change in vegetation can then affect the types and numbers of animals that depend on those plants for food and shelter.

Therefore, rising sea levels will impact both the population sizes and the variety of organisms that interact in the low marsh ecosystem.

25. **(3)** Planting vegetation supports ecosystem stability by holding soil in place, reducing erosion, and providing habitat for a wide range of organisms. This encourages complex interactions among species and a resilient food web.

Wrong Choices Explained

(1) One species of plant limits biodiversity.

(2) Bulkheads separate land and water, disrupting ecosystems.

(4) Revetments increase wave energy impact and reduce habitat quality.

26. **(2)** Oyster reefs reduce storm wave energy, which prevents erosion and allows salt marshes to migrate inland. This expands habitats for marsh organisms and helps maintain biodiversity.

Wrong Choices Explained

(1) Erosion would not reduce salt exposure—it increases it.

(3) Bulkheads prevent marsh migration, harming ecosystems.

(4) Storms do affect salinity, and reduced vegetation causes more erosion.

27. **Sample Response:** The best option for shoreline restoration in Oswego, NY is the revetment. A revetment lays stone, rubble, or concrete along the slope of the shoreline, making it highly reliable in protecting property from erosion and waves, which is important after the 2019 record high water levels. Although the initial construction cost is high ($$$$), it is low maintenance over time, which helps balance long-term expenses. Revetments also have the advantage of reducing wave action and protecting homes and businesses. However, they can fragment habitat and reduce the natural connection between land and water. Socially, this solution protects property and recreational areas, but aesthetically it looks less natural than vegetation or sills. Overall, the reliability of protection and long-term cost savings outweigh the environmental trade-offs, making revetments the best option.

28. **(2)** Similarities in the genetic sequences controlling limb development in zebrafish and tetrapods show a pattern of evolution from common ancestors. These shared genes indicate a continuity in body structure development from ancestral bony fish.

Wrong Choices Explained

(1) Habitat changes do not cause changes in gene sequences.

(3) Counting limbs does not provide genetic evidence.

(4) Usage of limbs does not show evolutionary origin.

29. **(4)** The diagram shows that both extinct and living species have forelimbs with a single upper arm bone, two lower arm bones, and multiple wrist and digit bones. These structural similarities are evidence of descent from a common ancestor.

Wrong Choices Explained

(1) *Acanthostega* lived in water, not on land.

(2) Habitat similarity does not prove closest relation.

(3) Forelimb differences are adaptations, not the result of separate evolution.

30. **Sample Response:** The fossil and structural evidence shows that tetrapod front limbs share a common pattern of bones, such as one bone in the upper arm, two bones in the lower arm, and many smaller bones in the wrist and digits. Even though these limbs are adapted for different functions in modern species—such as flying in bats, swimming in porpoises, or digging in moles—the same basic bone arrangement is present. This suggests that tetrapod front limbs evolved from a common ancestral structure found in ancestral bony fish, supporting the idea of descent with modification.

31. **Sample Response:** *Tiktaalik*'s front limbs show a combination of features found in both ray-finned fish and early tetrapods. Like fish, it still has fin rays made up of thin bones. However, it also has thick bones at the base of the fin that resemble the upper arm and forearm bones of tetrapods. This mixture of structures supports the idea that *Tiktaalik* was a transitional form, bridging the evolutionary gap between ancestral fish and the earliest four-limbed vertebrates such as *Acanthostega*.

32. **(4)** *Tiktaalik*'s limb bones helped it push onto land, where it could access food and avoid aquatic predators like *Rhizodus*. These traits improved survival and reproduction, showing that environmental pressures influenced evolution.

Wrong Choices Explained

(1) *Rhizodus*'s mobility is not linked to *Tiktaalik*'s land traits.

(2) Eye location is helpful but not the primary adaptive trait.

(3) Fangs help *Rhizodus*, not *Tiktaalik*.

33. **(1)** Carbon is absorbed by plants from the atmosphere and used to create sugars through photosynthesis. When animals eat plants, they use the sugars in cellular respiration to release energy, showing how carbon and energy flow through the ecosystem.

Wrong Choices Explained

(2) Oxygen is not used to make nutrients.

(3) Cows do not release sugars into the air.

(4) Wastes do not release sugars used by plants.

34. **Sample Response:** Process *C* represents the breakdown of dead organisms and wastes by decomposers. During this process, carbon and nutrients are released back into the soil and the atmosphere. These recycled nutrients are then available for plants to absorb and use during photosynthesis (Process *A*). This cycling ensures that matter continues to move between organisms in the ecosystem, supporting both plant growth and the animals that depend on plants for food.

35. **(3)** The elements in Substance 1 (likely carbon, hydrogen, and oxygen) combine with nitrogen to form proteins. This shows how plants rearrange absorbed materials to create essential organic molecules.

Wrong Choices Explained

(1) Nitrogen and phosphorus are not broken down from Box *Y* elements.

(2) Starch is formed from sugars, not from Substance 1 directly.

(4) DNA requires phosphorus and nitrogen, not only Substance 1.

36. **(2)** Photosynthesis (Process *A*) converts carbon dioxide and water into glucose (Substance 1), which is then used in metabolism. This shows how plants use inorganic carbon to produce energy-storing molecules.

Wrong Choices Explained

(1) Lipids are not the first product of Process *B*.

(3) Nitrogen is not combined directly in these processes.

(4) Proteins and DNA are made from different combinations of atoms.

37. **Sample Response:** Plants take in carbon dioxide from the atmosphere during photosynthesis, and this carbon is converted into organic compounds such as starch, lipids, proteins, and DNA. For example, the model shows that 123×10^{15} grams of matter are absorbed by plants from the atmosphere, but

only 60 × 10^{15} grams are later emitted back. This means that a large amount of carbon remains stored in plant biomass, reducing the amount of carbon in the atmosphere while increasing carbon stored in the biosphere. Thus, the cycling of matter by plants directly changes carbon levels between the atmosphere and living organisms.

38. **(3)** As the number of prairie dog burrows decreased, the population of burrowing owls also declined. This shows that prairie dog burrows serve as critical nesting sites and their availability affects owl carrying capacity.

Wrong Choices Explained

(1) Owl population decreased with fewer burrows.

(2) Owls are not a limiting factor for prairie dogs.

(4) Owl increase did not reduce burrow availability.

39. **Sample Response:** A significant decrease in the prairie dog population would reduce the number of burrows they create and maintain. Since abandoned prairie dog burrows provide essential shelter for species such as burrowing owls, rattlesnakes, and insects, fewer prairie dogs would mean fewer available habitats for these organisms. Additionally, prairie dogs are a primary prey species for animals like the black-footed ferret, so their decline would also reduce food sources for predators. This shows that the loss of prairie dogs would disrupt multiple interactions in the ecosystem, leading to widespread effects.

40. **(2)** The graph shows a population rebound in Colorado after a drop, suggesting some prairie dogs had a heritable trait that made them resistant to plague. These individuals survived and reproduced, increasing the resistant population.

Wrong Choices Explained

(1) Recovery came after survival of resistant individuals.

(3) Percent survival was not always high.

(4) South Dakota had no plague; both populations did not recover together.

41. **(3)** Preservation of grassland supports consistent food supply and habitat for prairie dogs. Predators like ferrets keep population growth in check, preventing overuse of resources.

Wrong Choices Explained

(1) Urban development reduces habitat.

(2) Depletion of soil and rangeland reduces capacity.

(4) Increased nutrients would raise population, not limit it.

42. **Sample Response:** The better option for protecting prairie dogs from the plague in this situation is vaccination. Although it requires prairie dogs to eat the tablets within 7 days, it is safer because it only affects the prairie dogs themselves and does not introduce chemicals into the environment. Vaccination fights off infection for up to 9 months, reducing the chance of disease spread while avoiding harmful effects on other animals, livestock, or people living near the area. Burrow dusting, while reliable at killing fleas, introduces insecticide powder into the soil, which could negatively impact other species and the surrounding ecosystem. Therefore, vaccination is the more environmentally safe and reliable method.

43. **(1)** To understand inheritance, it's important to determine whether genes related to birth weight were inherited from both parents. DNA from both parents influences offspring traits, including those seen during the famine.

Wrong Choices Explained

(2) Genes are made of DNA, not amino acids.

(3) Genes are not made of protein.

(4) Stomach cells are not involved in inheritance.

44. **(1)** Changing CTC to CTG results in a different mRNA codon, which changes the amino acid from glutamic acid to aspartic acid. This alters the protein and may affect its function.

Wrong Choices Explained

(2) The change does affect the protein.

(3) Valine is not the correct substitution.

(4) One amino acid changes—not all.

45. **Sample Response:** When DNA is heavily methylated, the added methyl groups block transcription, so the IGF2 gene cannot be expressed. In contrast, unmethylated DNA is less tightly wound and allows transcription to occur. Because the children of the famine had less methylation of the IGF2 gene, more of the gene was transcribed, producing greater amounts of the IGF2 hormone. Since IGF2 promotes fetal growth, this increase in hormone production led to higher birth weights, even though the mothers experienced poor nutrition.

46. **(2)** Similar health problems in offspring of famine-exposed individuals suggest that methylation patterns can be inherited across generations, affecting gene expression without changing DNA sequence.

Wrong Choices Explained

(1) Methylation was passed beyond one generation.

(3) Gene expression is also affected by epigenetics.

(4) Methylation doesn't prevent all mutation.

47. **Sample Response:** The model shows that as DNA becomes increasingly demethylated, control over gene expression is disrupted. In lung cells, this results in abnormal activation of genes that regulate cell division. Without normal regulation, the cells begin dividing uncontrollably, forming tumors that progress from early to intermediate to late stages. Thus, the disruption of the normal flow of genetic information caused by DNA demethylation leads to uncontrolled cell growth and cancer development in lung cells.

48. **(3)** Using medication to add methyl groups can suppress genes involved in uncontrolled cell division, reducing tumor growth. This targets the source of the problem—unregulated gene expression.

Wrong Choices Explained

(1) Radiation removes methyl groups, which could worsen cancer.

(2) Accelerating cell division in tumors is harmful.

(4) Increasing mitosis could promote tumor growth.

Regents Practice Exam August 2025

***Directions:* For each multiple-choice question, record in the space provided the number of the choice that best completes the statement or answers the question. For questions requiring a written response, write your answer in the spaces provided.**

Base your answers to questions 1 through 4 on the information below and on your knowledge of biology.

Trees Have Organ Systems

Trees have two systems, compared to the multiple systems that are present in animals. Trees are multicellular organisms with organ systems that enable them to carry out specific functions necessary for maintaining homeostasis. However, that does not diminish the importance of these two systems, which have numerous critical functions.

The table below contains information about the structures and functions of the systems found in trees.

Systems in Trees

	Root System	Shoot System
Structures	Roots	Stem/trunk, branches, leaves
Functions	- Anchor the tree - Absorb water and minerals from the soil - Store and modify products of photosynthesis	- Connect roots to branches - Transport material - Perform photosynthesis - Contain reproductive structures

The model below is a cross section of a stem as viewed through a microscope that was observed during an investigation. The stem contains the xylem and phloem, which together make up the vascular bundle, a structure that transports materials within a tree.

Cross Section of a Stem

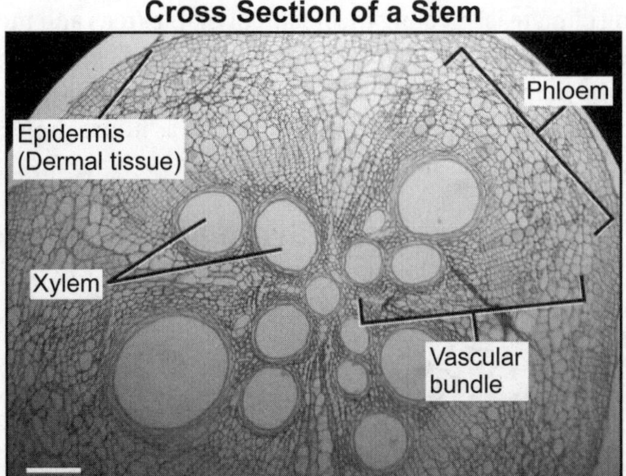

1. Describe an interaction between the vascular bundle with both the root and shoot systems in a plant. [1]

____

____

____

The compounds needed by plants to carry out life functions are delivered to plant cells in different ways. Maple trees are one type of tree that produce a liquid, known as sap, that circulates sugar, water, and other molecules throughout the plant. Root hairs are single-celled structures that exchange oxygen and carbon dioxide within air pockets in the soil. They also absorb water from the soil. These materials are then delivered to all plant cells.

2. Which statement best explains why the function of sap and root hairs is necessary for the functioning of plant cells to maintain homeostasis?

(1) Cellular respiration occurs only in root cells since the sap only travels to those cells.

(2) The sugar in the sap combines with carbon dioxide to form new sugars for the root cells.

(3) Sap and root hairs contribute to the transport of materials necessary for the plant cells to carry out cellular respiration.

(4) Sap and root hairs contain the raw materials required for the root cells to carry out photosynthesis.

2 ______

Changing climate is affecting the health of maple trees and their production of sap, used to produce maple syrup. The result is less sweet syrup, which is affecting the maple syrup industry. The model below shows the amount of maple syrup produced from 1950 to 1999, as well as future projections, given the continual rise in temperature.

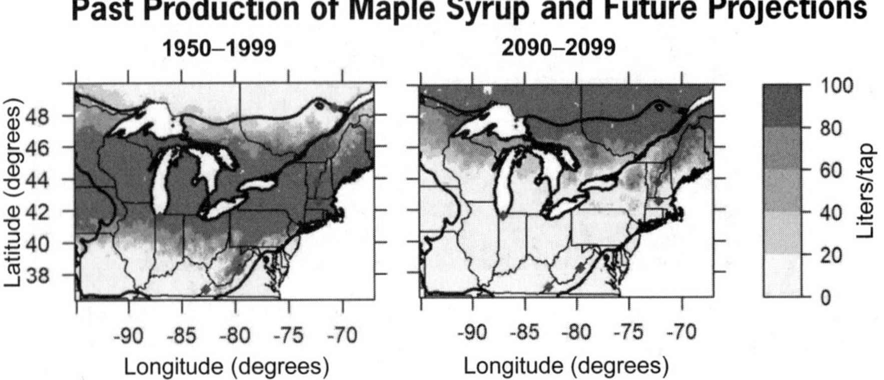

3. Which statement provides evidence to explain how the changing climate will affect maple trees and syrup production in the future?

(1) The optimal areas for maple syrup production will shift northward where temperatures are cooler.

(2) The optimal areas where maple syrup was produced from 1950 to 1999 will stay the same in the future.

(3) Future maple syrup production will mainly occur in latitudes between 38 and 42 degrees, since maple trees will be more successful in that range.

(4) Maple trees will not survive in northern areas in the future; therefore, maple syrup production will stop.

3 ______

Scientists are concerned about the changing climate and the potential loss of maple trees. Trees are extremely important in the sequestering (storage) of carbon.

The models below illustrate carbon sequestration based on tree age and the diameter of various species of trees.

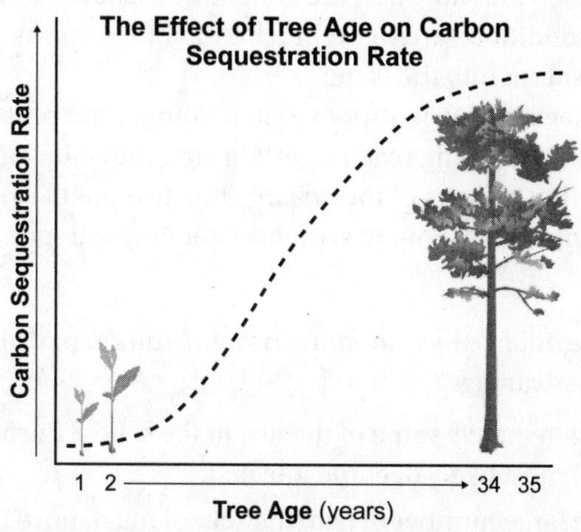

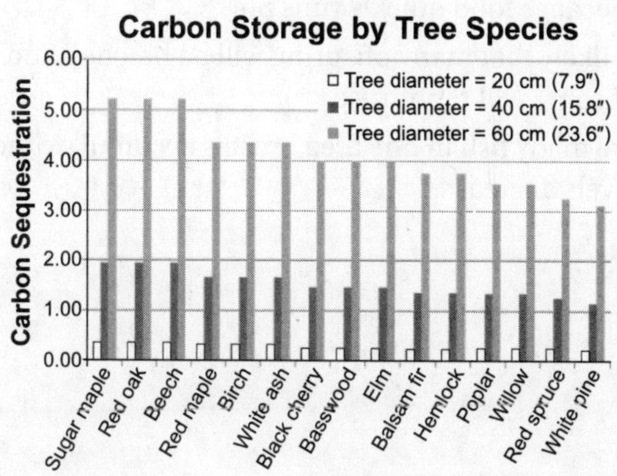

4. Describe the role of tree age *and* diameter, including those of maples, on carbon cycling and storage between the atmosphere and biosphere. [1]

____

____

____

____

Base your answers to questions 5 through 10 on the information below and on your knowledge of biology.

Strategies for Survival in Different Animal Species

When bluefin tuna reach breeding age, they migrate in schools until they reach an area in the ocean where conditions are right. Bluefin tuna then spawn, where they release millions of eggs and sperm into the water.

Spawning takes place in nutrient-poor water. Young tuna practice cannibalism to obtain nutrients. Once the young reach a certain size, they stop eating each other and travel in schools to other regions of the ocean. Only two out of every 30 million fertilized eggs will reach adulthood. Four to six years later they will spawn, and the cycle will begin all over again.

5. Which piece of evidence best supports how the tuna's spawning behavior provides a survival advantage?

(1) If a pathogen infects some of the fish in the school during spawning, then others will most likely become infected.

(2) Due to the large number of fish, the school must move from one area to another because food quickly runs out.

(3) It is more likely that many offspring will be hatched, increasing the likelihood that some will reproduce.

(4) If there are many fish in one area, wastes accumulate more quickly, and oxygen levels decrease.

5 ______

Shrews are mammals that live primarily underground. The female gives birth to five to seven offspring per litter and has three or four litters per year. About 50% of the offspring survive. The young are dependent on the mother for milk for 22 to 25 days. During this time, if the mother has to move the nest to a new location, the young shrews form a caravan behind the mother. Each shrew holds onto the base of the tail of the shrew in front of it with its mouth, forming a chain of shrews, as shown below.

6. Construct an explanation that shrew behavior makes it possible for the species to survive even though they produce fewer offspring than the tuna per reproductive cycle. [1]

____

____

____

A survivorship curve is a model showing the number or proportion of individuals surviving to each age for a given species or group. The survivorship curves below are used to represent life expectancy patterns in three different species—*A*, *B*, and *C*.

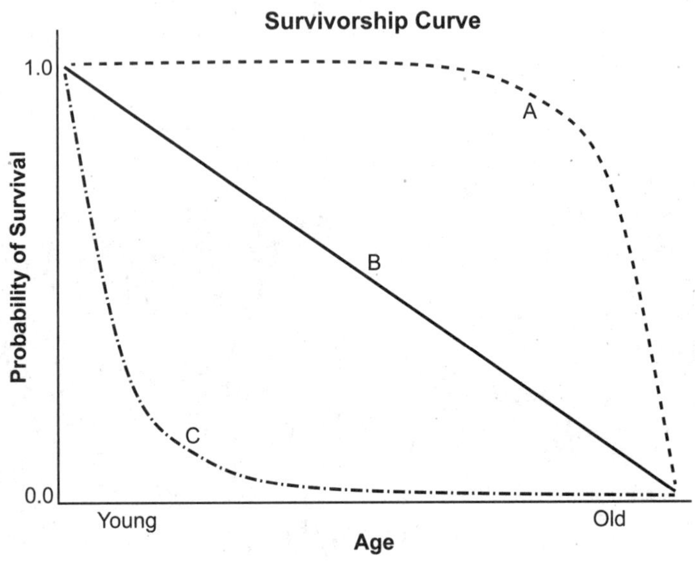

7. A student claimed that *C* best represents the survivorship curve of a bluefin tuna. Which factor explains this claim?

(1) Millions of eggs and sperm are released at the same time in the same area of the ocean.

(2) An extremely small number of the young that hatch survive to adulthood.

(3) The young migrate to a different region of the ocean upon reaching a certain size.

(4) Because they practice cannibalism, approximately half of the young reach adulthood.

7 ______

8. Use the information provided to explain which survivorship curve best represents the shrews. [1]

______

______

______

American bison are herding mammals found on the plains of North America. When predators, such as wolves or mountain lions, threaten the calves, the adults form a double ring around the calves. The adult female bison make an inner circle around the calves while the male adult bison form an outer circle around the females.

A Herd of American Bison

9. Explain how the process of evolution has led to the development of this behavior in bison populations. [1]

____

____

____

10. In the 1800s, unregulated hunting reduced bison herds from tens of millions of individuals to fewer than 1,000. How might the DNA of the modern bison populations compare with the DNA of ancestral populations prior to the 1800s?

(1) Modern bison have more genetic diversity than their ancestors as a result of an increase in available mates.

(2) Modern bison have less genetic diversity than the population before the 1800s did as a result of increased random mutations.

(3) The DNA of the modern bison is identical to that of bison living before the 1800s because they all evolved from the same common ancestor.

(4) The DNA of modern bison shows less genetic diversity compared to their ancestors due to the limited genetic variation remaining in their population.

10 ______

Base your answers to questions 11 through 14 on the information below and on your knowledge of biology.

A Unique Relationship: Mulberries and Silk

The ability of living organisms to rearrange elements into different forms and groups is fundamental to life. Mulberry plants use specific biological processes to combine molecules present in the environment in order to produce the substances they need to carry out life functions. Silkworms eat mulberry leaves exclusively, getting all of their water and other nutrients from these leaves.

Most nutrients from the mulberry leaves enter the silkworm in the form of macromolecules. The macromolecules must be metabolized by body systems of the silkworm into a form that the cells of the worm are able to use. The model below summarizes some of the processes the silkworm uses to convert the nutrient macromolecules from the mulberry tree into a usable form.

Possible Metabolic Pathways of Macromolecules in Silkworm

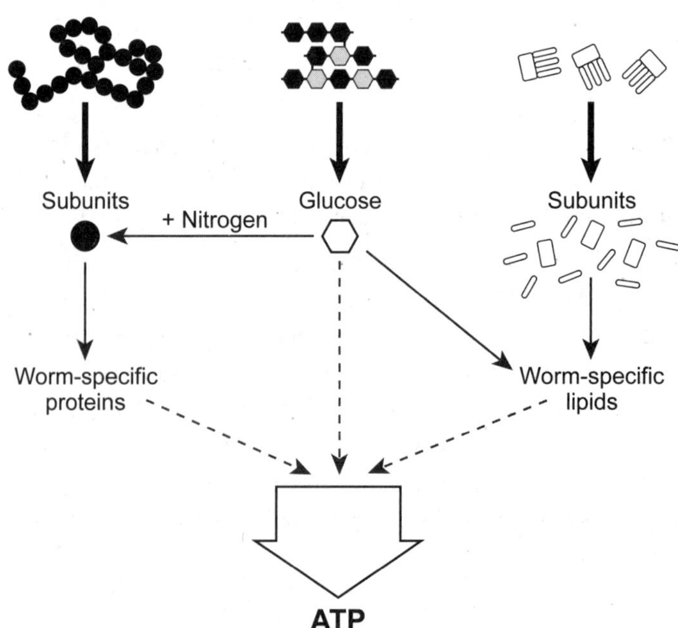

11. Based on information in the model, which statement describes the chemical process by which the energy in glucose from the mulberry plant is converted and made available to the silkworm?

(1) Cellular respiration within the cells of the silkworm breaks the bonds of glucose to release energy that can be used to produce ATP, a usable source of cellular energy.
(2) Cellular respiration within the cells of the mulberry tree breaks the bonds of glucose to release amino acids, a usable source of cellular energy.
(3) Digestion within the cells of the silkworm breaks the bonds of glucose to release amino acids that can be used to build ATP for energy.
(4) Digestion within the cells of the mulberry tree breaks the bonds of glucose to release elements that can be used to build lipids for energy.

11 ______

12. Use evidence from the model to explain how the elements in the glucose can be used by the silkworm to synthesize protein molecules needed to sustain life. [1]

____

____

____

Researchers have determined that silkworm moths have a mechanism that enables them to locate mulberry leaves. Mulberry leaves emit scented chemical molecules that are detected by highly sensitive receptors within the antennae of the silkworm, as shown in the models below.

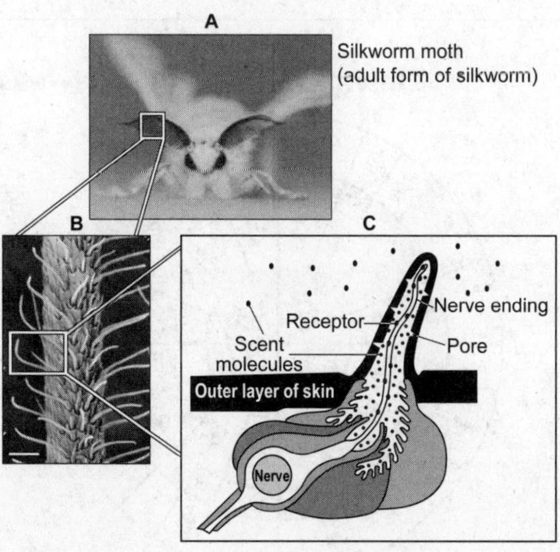

13. Which statement best describes how the organization and interactions of *two* systems present within the silkworm moth enables it to find and use mulberry leaves as their source of nutrients?

(1) Special receptors in the nervous system of the silkworm pick up the scent of the mulberry leaves and send messages to the muscular system to spin a silk cocoon.

(2) Receptors in the nervous system within the antennae pick up scent molecules from mulberry leaves. The silkworm is able to find and eat the leaves, using the digestive system to break down the nutrient molecules into a usable form.

(3) The nervous system of the silkworm sends messages to the digestive system to begin to break down the fats that are taken in to produce new muscle tissue for movement.

(4) The muscular system sends messages to the nervous system to receive scent messages from the mulberry leaves.

13 ______

Silkworms use the mulberry leaves to produce fibroin, a protein molecule needed to spin silk. Part of the DNA sequence that codes for the protein fibroin is listed in the table below.

14. Use the partial DNA code to determine the mRNA sequence. Then use the Universal Codon Chart below to determine the amino acid sequence that would result if this DNA sequence was expressed to produce fibroin. [1]

Partial DNA	GAT	CAA	TTA	AAT
mRNA				
Amino acids				

Universal Codon Chart

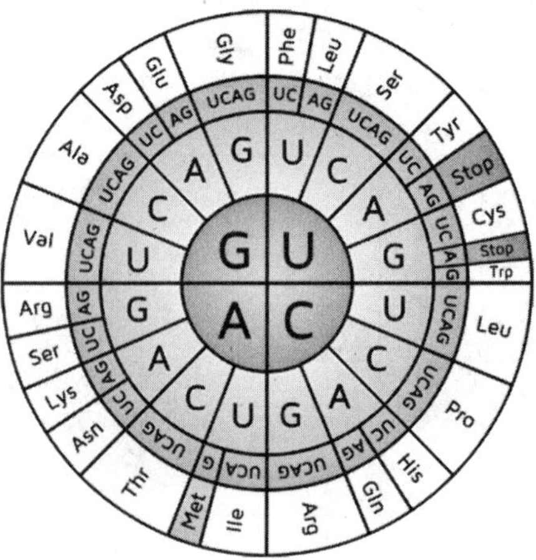

Base your answers to questions 15 through 19 on the information below and on your knowledge of biology.

Ivory and Elephants

In African elephant populations, both males and females usually have ivory tusks, which are actually a pair of massive teeth. Some females never grow them. Tusks are crucial to male elephants for defense and in their competition for mates.

In many parts of the world, elephant ivory is still viewed as a status symbol. It was used to produce small carvings, jewelry, piano keys, and chess sets. Traditional medicine used ivory powder to treat a variety of illnesses. Acquiring ivory results in the illegal killing of elephants, known as poaching. Ivory is still sold, even though it is against the law to do so.

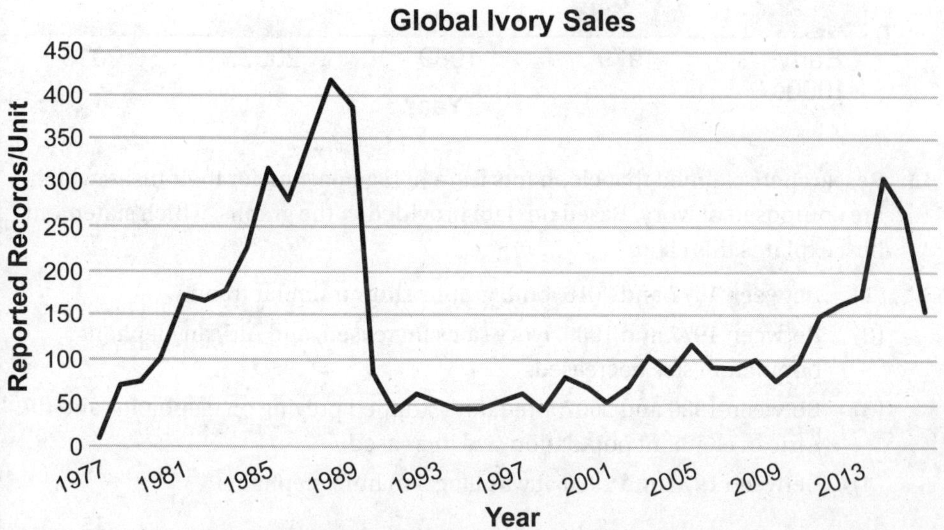

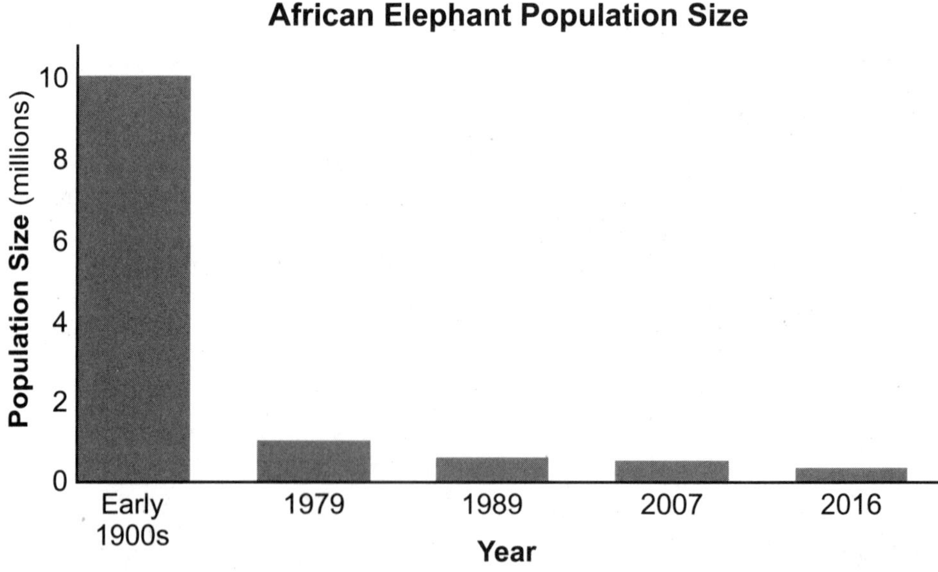

15. Researchers claimed that elephants were being poached for their tusks, which are composed of ivory. Based on data provided in the graphs, which statement best explains this claim?

(1) Between 1977 and 2016, both graphs show a similar trend.

(2) Between 1977 and 1989, ivory sales increased, and African elephant population size decreased.

(3) Between 1980 and 2007, predators stopped preying on elephants, and the African elephant population size increased.

(4) Between 1979 and 1990, it was illegal to hunt elephants.

15 ______

Tusklessness, not growing tusks, appears to be caused by a dominant allele, carried by females and lethal to males. The table below contains data on the change in the frequency of tusklessness in female African elephants.

Female African Elephant Tusklessness Data

Year	Sample Size	Tusked Females	Tuskless Females	Tusklessness %
1969	247	221	26	10.5
1972	205	180	25	12.2
1988	132	89	43	32.6
1989	165	102	63	38.2
1990	298	194	104	34.9
1991	171	114	57	33.3
1992	195	136	59	30.2
1993	279	199	80	27.7

16. Which statement best describes a trend in tusklessness represented in the data table?

(1) From 1988 to 1991 the frequency of tusklessness decreased.
(2) From 1969 to 1989 the frequency of tusklessness increased.
(3) From 1969 to 1993 the total number of tuskless females remained constant.
(4) From 1989 to 1991 the total number of tuskless females increased.

16 ______

17. Which statement about how humans affected the elephant population is supported by the evidence provided?

(1) The decrease in global ivory sales resulted in a decrease in tuskless male elephants.
(2) The chance of producing tuskless males increased as a result of the tighter enforcement on poachers.
(3) Poaching has led to an increase in the tuskless trait among female elephants.
(4) The demand for ivory has led to a decrease in the number of tuskless female elephants producing offspring.

17 ______

African savanna elephants can significantly impact the ecosystems that they live in. These elephants can push down and uproot trees, allowing the tree bark to be used for food. As trees are uprooted, the soil is disturbed, and the organisms that live under the soil are exposed.

The table below shows how the trait of tusklessness impacts the behavioral characteristics of African savanna elephants.

Behavior	Tusked Elephants	Tuskless Elephants
Defense	Use tusks as primary defense	Rely on strength and group cooperation
Hole-digging and use	Can dig deep holes in compacted soil to access water and minerals	Mainly use existing holes to access water and minerals
Dominance vocalization	Deep rumbles and trumpeting	Deep rumbles and trumpeting
Feeding	Eat more fibrous and woody vegetation	Eat more grasses and soft vegetation
Knocking down trees	Can use both tusks and trunk	Can use trunk only

18. Which statement identifies the claim, supported by evidence, that a shift to more tuskless elephants could disrupt the stability of the local ecosystem?

(1) The amount of grass available to other herbivores in the ecosystem would decrease.

(2) There would be more deep watering holes available within the ecosystem.

(3) The amount of living woody vegetation available to organisms in the ecosystem would decrease.

(4) Soil-dwelling organisms would be more vulnerable to predators, and their number would decline.

18 ______

Conflict between humans and elephants exists across Africa. As human populations grow, people are moving into wild areas, increasing loss of natural habitats. Elephants compete with people for decreasing resources. Elephants searching for food frequently enter villages and sometimes damage property, uproot vegetable gardens, and can harm people. Multiple organizations have come up with solutions to reduce conflict between elephants and humans, as described below.

Solution One: Beehive Fences

Beehive fences are made by hanging beehives connected by wire around crops to deter elephants from entering areas by using their natural fear of African honeybees. When an elephant pushes the wire the hives shake, alerting guard bees which then defend their hives. Over time, elephants learn to avoid these bee-fenced areas.

Solution Two: Resource Management Hunting

Millions of acres of elephant habitat are set aside where limited elephant hunting is allowed. These areas are closely monitored and regulated. Without this managed hunting, the critically important elephant range will be lost to other land uses.

19. Evaluate the solutions listed above by explaining which solution would be most effective at reducing the conflict between humans and elephants while having the *least* impact on the ecosystem. [1]

____

____

____

Base your answers to questions 20 through 23 on the information below and on your knowledge of biology.

Carbon Cycle Supports Life on Earth

Life on Earth is only possible with carbon. Organisms on Earth rely on carbon for the production of molecules used in life functions. Various processes in the environment cycle carbon. In the ocean, the main producers in this cycle are plant-like microscopic organisms called phytoplankton.

20. Which model below best identifies how phytoplankton cycle carbon between two of Earth's spheres?

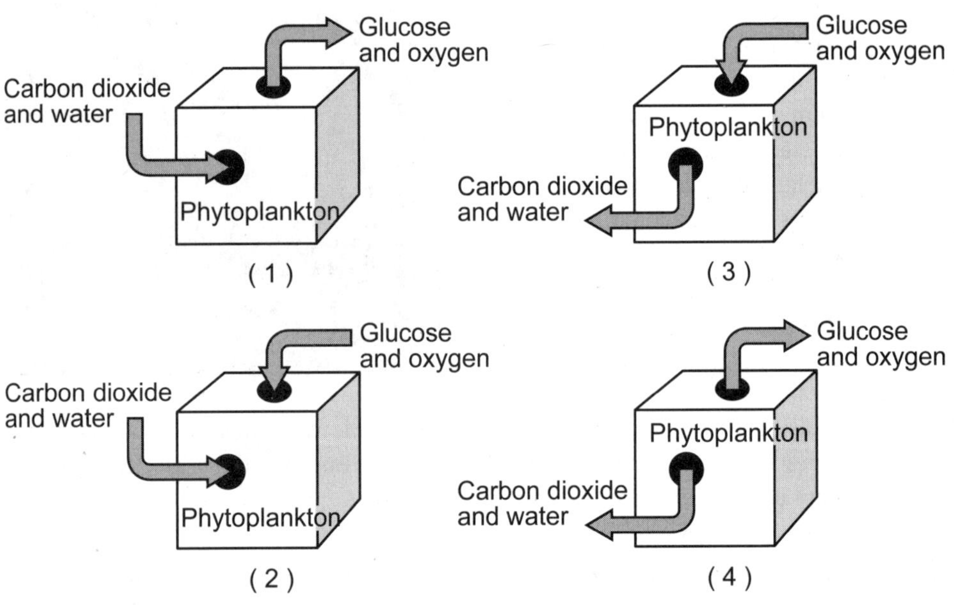

20 ______

The model below represents how processes *A*, *B*, and *X* play a significant role in the exchange of carbon.

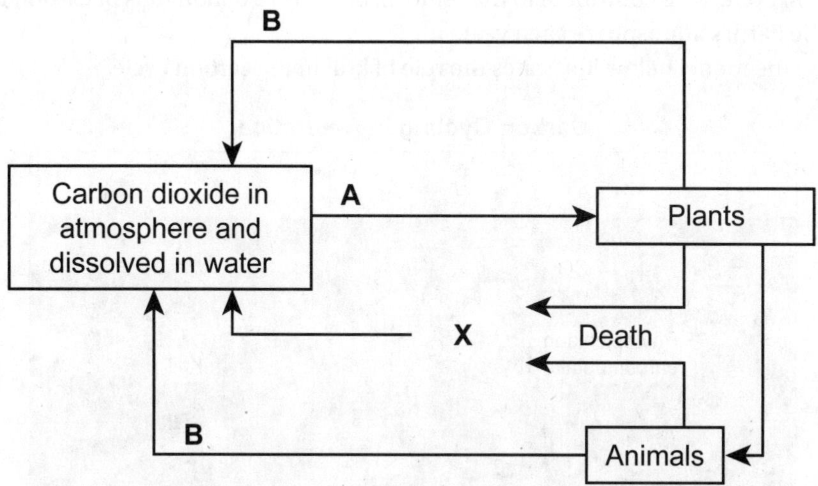

21. Based on the model, identify the process represented by *X and* describe how this process contributes to the cycling of carbon between *two* of Earth's spheres. [1]

____

____

____

Krill are small shrimp-like creatures that live in the southern oceans around Antarctica and consume phytoplankton. They play an important role in the carbon cycle. Krill contribute to the removal of up to 12 billion tons of carbon from the Earth's atmosphere each year.

The model below illustrates the role of krill in the carbon cycle.

Carbon Cycling in Antarctica

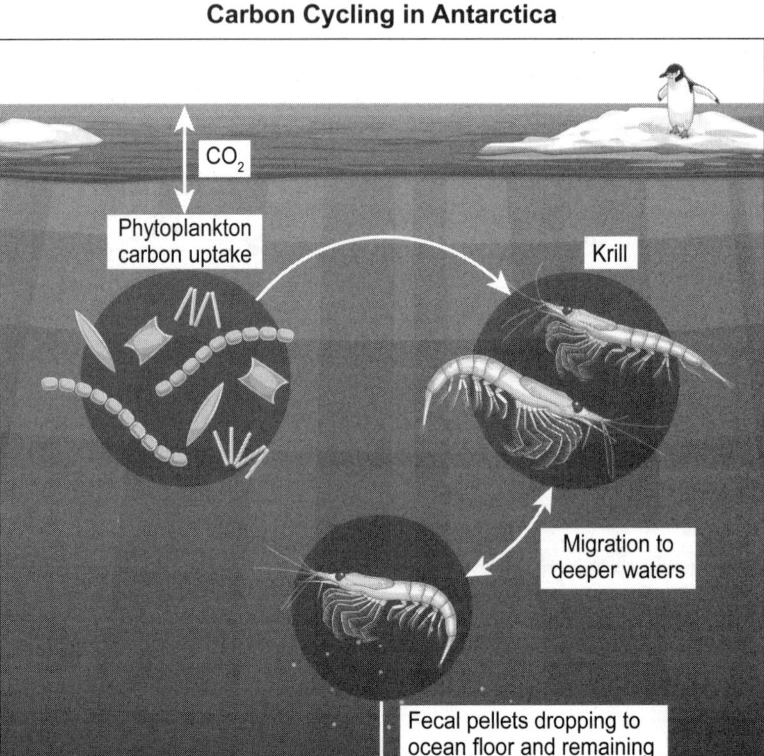

22. Based on the model, explain how rapid growth in the krill population would affect the cycling of carbon between the atmosphere and biosphere. [1]

____

____

____

23. A carbon sink is anything that absorbs more carbon from the environment than it releases. Which process carried out by krill results in the accumulation and storage of carbon for a period of time?

(1) Photosynthesis in the atmosphere releases carbon to the atmosphere.
(2) Respiration in the geosphere releases carbon wastes into the atmosphere.
(3) Elimination of wastes in the hydrosphere stores carbon on the ocean floor.
(4) Acidification in the biosphere stores carbon in the bodies of organisms.

23 ______

Base your answers to questions 24 through 28 on the information below and on your knowledge of biology.

Keratin—A Very Versatile Protein

The skin, hair, nails, horns, and claws of many organisms are composed of a tough, structural protein called keratin. In humans, there are over 54 different types of keratin proteins in the body. The KRT1 gene, involved in making one type of keratin protein called keratin 1, is expressed in the outer layers of skin. Another keratin protein is coded for by the gene KRT12 and functions in the cornea of the eye.

Segments of the DNA code for synthesizing these proteins are shown in the table below.

Gene	DNA Sequence							
KRT1	TAC	AGA	GGA	GTG	TTT	AGC	TCC	TTC
KRT12	TAC	TTC	AAT	CCC	TAT	TGT	CTG	TTC

24. Which statement best provides evidence that supports the explanation that the sequence of DNA determines proteins?

(1) The structure of KRT1 and KRT12 mRNA determines the DNA code that is translated to produce a specific protein with a specific shape, which determines its function.

(2) The structure of the KRT1 and KRT12 genes determines the genetic information present in protein that determines the function of each DNA molecule.

(3) The DNA that codes for KRT1 and KRT12 is different, resulting in different protein structures with different functions.

(4) The structure of each protein determines the order of bases in DNA. The KRT1 and KRT12 proteins are different, resulting in different functions.

24 ______

The model below represents the approximate location of the human KRT1 and KRT12 genes within the human genome.

Chromosomes Present in Human Cells

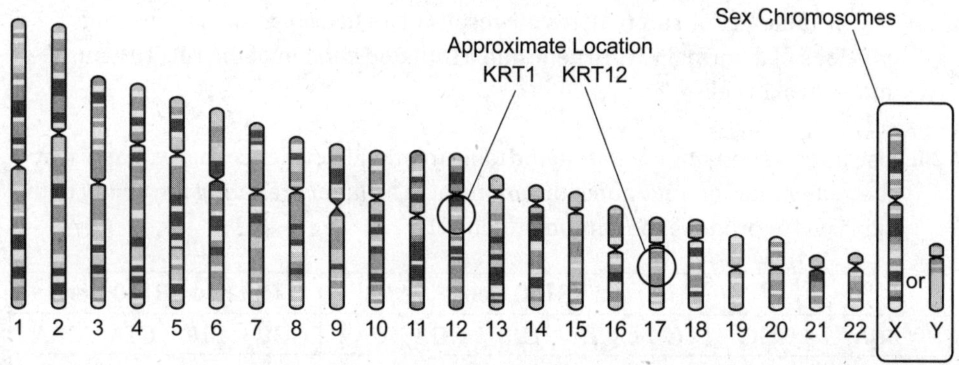

25. Use the model to describe how this genetic information is utilized to produce specialized cells. [1]

____

____

____

Some keratin-producing genes, like KRT10, are expressed in a group of body cells, known as keratinocytes. These are located in the outermost layers of the skin, making up about 90 percent of the cells there. In some individuals, keratinocytes do not produce appropriate amounts of keratin, resulting in various skin disorders.

The table below summarizes the results of an investigation used to study portions of a normal KRT10 gene and a mutated version of the KRT10 gene found in skin cells.

26. Use information obtained from the table to provide evidence that explains why the gene mutation may affect the amino acid sequence, altering the ability of the skin cells to produce the keratin protein. [1]

	Normal KRT10 Gene	Mutated KRT10 Gene
DNA	GGC TTC CTA CTT GGA CAA	GGC TTC CTA CAA
Amino acid sequence	Pro Leu Asp Glu Pro Val	Pro Leu Asp Val

____

____

____

Treatments for some keratin-related disorders are being researched, including gene therapy. One possible technique would use lab-grown, genetically modified keratinocytes. These modified cells include a properly functioning version of the keratin-producing gene that is then transplanted into the patient.

27. Which claim best explains why only one generation of genetically modified keratinocytes must be genetically altered and transplanted into an individual in order for that individual to continue producing appropriate amounts of keratin?

(1) The lab-grown keratinocytes will undergo meiosis and pass the keratin DNA to future generations of skin cells.

(2) The modified cells will undergo mitosis and pass the functioning keratin gene on to future generations of skin cells.

(3) Only the modified skin cells will be able undergo mitosis, causing all body cells of the individual to produce appropriate levels of keratin.

(4) When the cells undergo meiosis, the DNA will be modified in future generations of skin cells to include only the genes for appropriate levels of keratin production.

27 ______

Crocodile scales, cat claws, and fur are also composed of keratin. The keratin in each of these structures is not identical, nor does it perform identical functions.

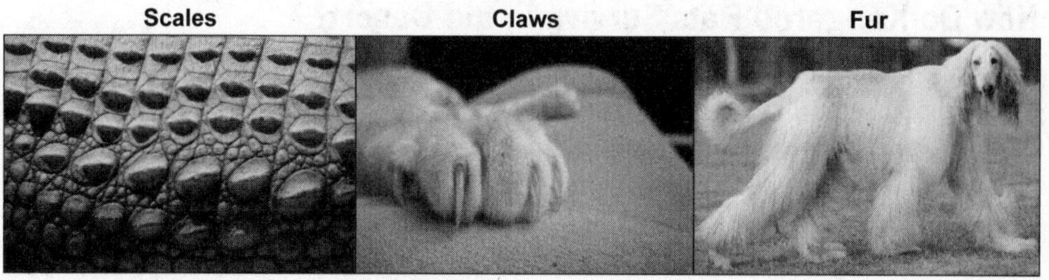

28. Based on the evidence provided, which statement best explains the variation in keratin structure and function as represented by these organisms?

(1) These organisms needed different structures to survive in their environments. The DNA that codes for keratin changed to improve each of these organisms' chance of survival.

(2) The gene responsible for the production of keratin in each organism had survivable changes to the DNA sequence, which were passed on to offspring.

(3) All of these organisms had ancestors that lived in the same environment and produced keratin that then mutated to produce different structures which were passed on to their offspring.

(4) Keratin most likely appeared first in crocodiles. The ancestors of the other organisms inherited the keratin gene from the crocodiles and modified it to adapt to their environment.

28 ______

Base your answers to questions 29 through 32 on the information below and on your knowledge of biology.

How Do Kangaroo Rats Survive in the Desert?

Kangaroo rat species live in the desert and conserve water so efficiently that they can survive without drinking. The body systems of these animals have various adaptations that allow the rats to extract water from the food they eat.

Kangaroo rats produce highly concentrated urine containing a minimal amount of water. Urine concentration depends on specialized structures in the kidney called nephrons.

The model below shows the organization of specific parts of the circulatory and excretory systems that are needed to filter water and waste from the blood.

Flow of Blood Through Kidney

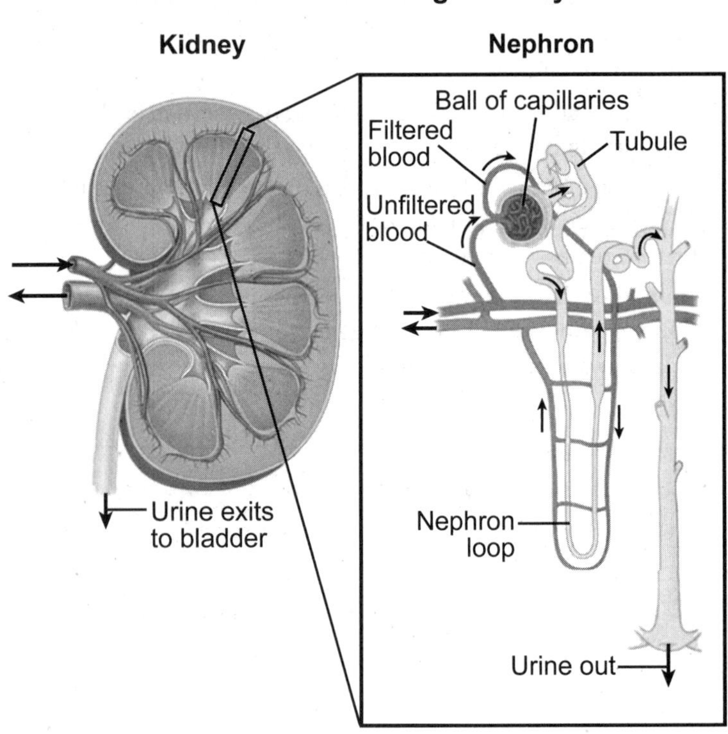

29. Describe how specific parts of the circulatory *and* excretory systems interact within the kangaroo rat to conserve water. [1]

____

____

____

The length of the nephron loop has been directly correlated with the efficiency of water conservation in organisms.

Location and Various Lengths of Nephron Loops

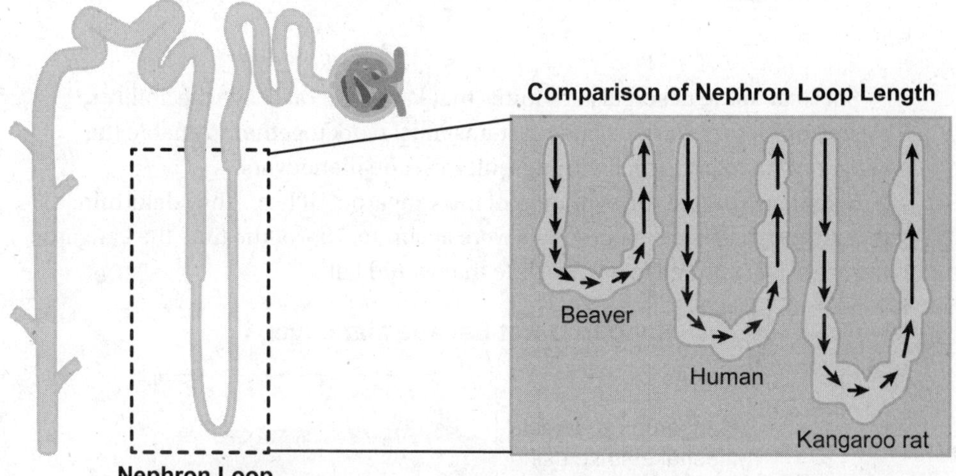

Urine Concentration in Mammals Living in Different Environments

Mammal Species	Environment	Maximal Urine Concentration (Milliosmoles/L)	Urine:Blood Plasma Ratio
Beaver	Fresh water	520	1.7:1
Human	Land with moderate moisture	1400	4.5:1
Kangaroo rat	Desert with low moisture	5500	16.0:1

30. Which explanation, based on the evidence provided, supports the claim that natural selection leads to the adaptations that control water regulation in specific environments?

(1) The shorter nephron loop in humans produces urine with the lowest urine:blood plasma ratio.

(2) The kangaroo rat has a long nephron loop and produces urine with the highest urine:blood plasma ratio.

(3) There is no relationship between the nephron length and maximal urine concentration.

(4) Mammalian kidneys have nephron loops that are the same length regardless of their environment.

30 ______

Survival in the desert also requires that kangaroo rats have the ability to escape predators. Various body systems must work together to enable the kangaroo rat to perform lightning-quick escape maneuvers.

Scientists studied the evolution of the kangaroo rat leap. They determined that although 81% of snake strikes were accurate, 78% of the time the kangaroo rats were fast enough to evade a bite that would kill.

Kangaroo Rat Escape Maneuver

Kangaroo rat detects rattlesnake strike using hearing, vision, or sensing ground vibrations. The kangaroo rat will initiate escape response.

Kangaroo rat completes escape maneuver by launching itself into the air.

31. Which statement best describes an interaction that occurs between body systems during the kangaroo rat's lightning-quick maneuver?

(1) The respiratory and nervous systems interact when sending a signal to the spinal cord to leap.

(2) The circulatory and nervous systems interact when sending a signal to the leg muscles.

(3) The respiratory and muscular systems interact to slow cellular respiration prior to the muscle contraction.

(4) The nervous and muscular systems interact to trigger the leg muscle to contract.

31 ______

California is home to several species of kangaroo rats. Kangaroo rats are highly affected by changes in their habitat. Because they don't drink water, they are dependent on the food available in the habitat for both nutrition and water.

The graph below shows the change in drought conditions in California over a 16-year period.

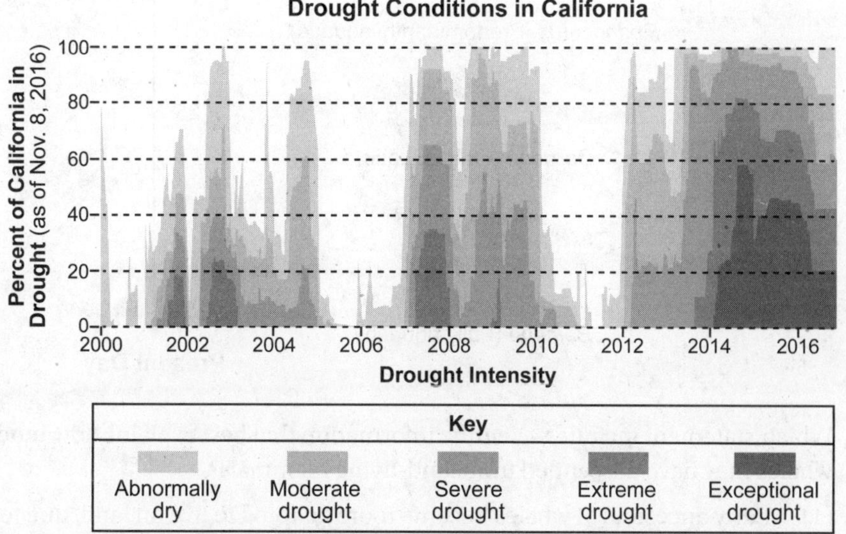

32. Using evidence from the information provided, describe how the kangaroo rat carrying capacity may be impacted if this trend continues in California. [1]

____

____

____

Base your answers to questions 33 through 36 on the information below and on your knowledge of biology.

Evolutionary Relationships

Evidence suggests that whales have descended from mammals that walked on land. The diagrams below show some information about extinct ancestors of modern whales.

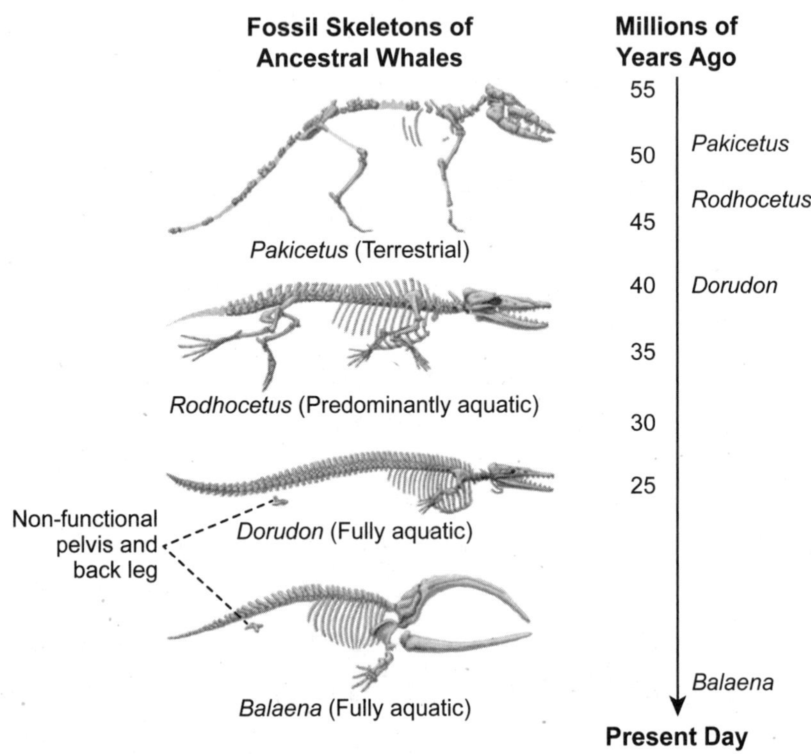

33. Which statement includes scientific information that best explains how modern whales may have descended from land-living mammals?

(1) Early ancestors of whales became more adapted to live on land, due to their paddle-like feet.

(2) The total number of whale species with a non-functional pelvis increased on land and in water.

(3) Ancestors of modern whales had leg and pelvic bones, which are present in more modern whales.

(4) Some ancestral whales learned aquatic behaviors that could be passed on to offspring.

33 ______

Whales and other marine mammals are often mistaken for fish, such as sharks, because of the similarity of their body shapes, as shown below. Even though they look similar, whales and sharks are genetically different.

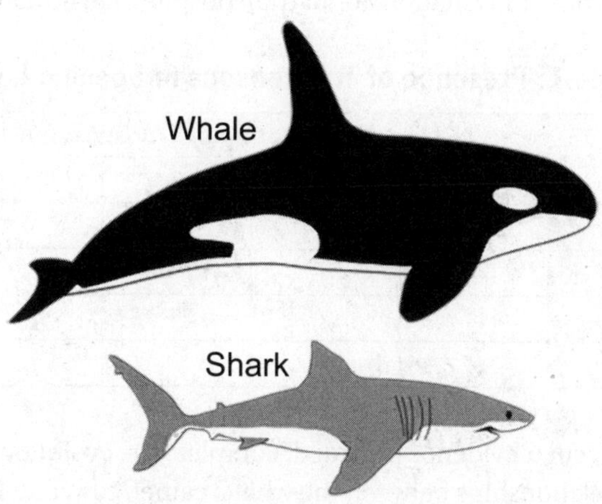

34. Based on evidence, which statement explains how whales and sharks have evolved to have similar body shapes?

(1) All aquatic organisms are closely related so they are likely to look the same.

(2) Marine mammals evolved directly from fish and therefore share many structural similarities.

(3) Whales and sharks possess identical genetic mutations in their genome, which determine their similar body shape.

(4) The body shape of whales and sharks allows them to swim efficiently, which improves their chances of survival.

34 ______

Researchers also studied the location of transposons, a type of DNA that is randomly inserted into a genome and passed down to future generations. Transposons are studied to determine evolutionary relationships. The table below shows some information about transposons in mammals.

Table 1: Presence of Transposons in Specific Locations

Animal	Location of Transposons				
	1	2	3	4	5
Whale	✓	✓	✓		✓
Camel					
Cow	✓	✓		✓	
Hippopotamus	✓	✓	✓		

35. Using the molecular evidence provided, complete the evolutionary tree below to describe the relationships between the whale, camel, cow, and hippopotamus. [1]

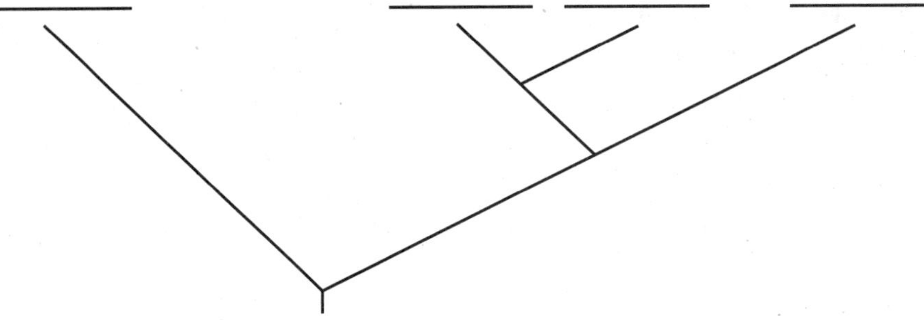

36. Which types of evidence would best demonstrate patterns that could be used to determine the evolutionary relationships between goats and the other organisms?

(1) Embryonic development, habitat, DNA sequences
(2) DNA sequences, location of transposons, skeletal structures
(3) Habitat, embryonic similarities, fossil record
(4) Leg bones, coloration, number of transposons

36 ______

Base your answers to questions 37 through 41 on the information below and on your knowledge of biology.

The Human Journey from One Cell to 30 Trillion

Events that occur during human reproduction and development result in the transformation of a single cell into an individual made of trillions of cells. These events are numerous and diverse, but all play a part in ensuring the continuity of life. The journey begins with the formation of the first cell (Process *X*). That cell will respond to specific factors both inside and outside the cell to produce many similar cells (Process *Y*).

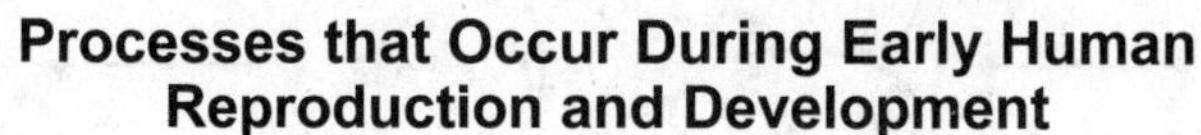

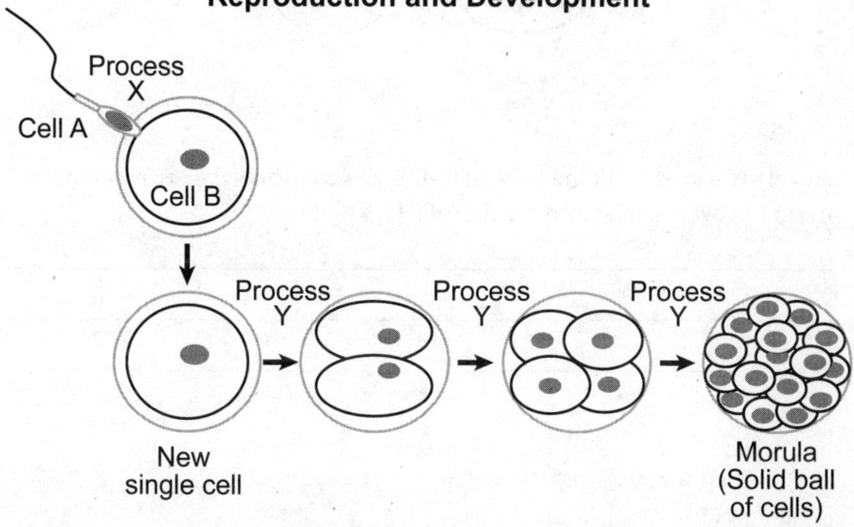

37. Which statement explains why Process *X* is necessary to maintain the continuity of life?

(1) It results in a cell that contains all of the genetic material needed to form an embryo.

(2) It guarantees the survival of the offspring in any kind of environment.

(3) It produces offspring with only traits that are favorable for survival.

(4) It assures that each parent passes on identical genetic material to the offspring.

37 ______

38. Which diagram best summarizes how Process *Y* results in the typical number of chromosomes present in the nucleus of each human cell?

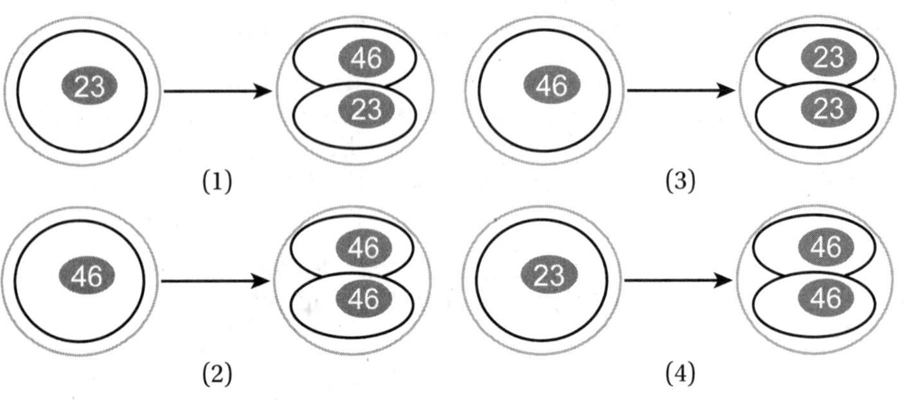

38 ______

39. If an error is present in the DNA of Cell *A*, describe how the error would be present in every cell within the morula. [1]

______

______

______

As human development continues, the solid ball of cells will hollow out to form a fluid-filled sphere of cells. At one end, it contains a mass of less than 100 cells called the inner cell mass (ICM). The ICM will form the fetus.

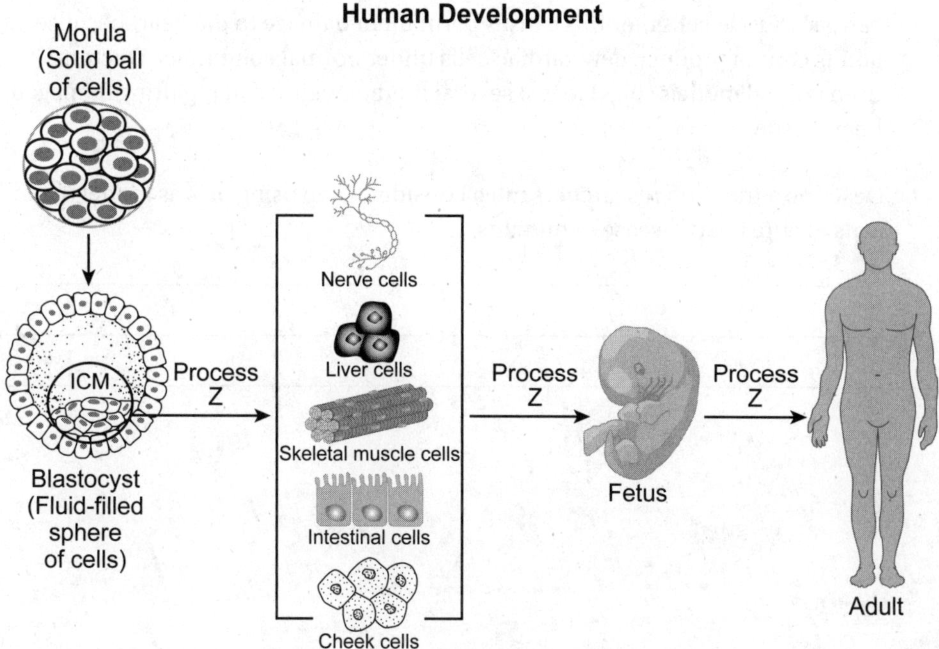

40. Which statement best summarizes the result of Process *Y* at all stages of human development, from the new single cell stage to the adult stage?

(1) Damaged genes are eliminated and favorable genes are passed on to the next generation of human cells.

(2) Cells are produced that can be used to produce offspring, which will increase the chance of the species surviving.

(3) The genetic variation between the cells of the developing organism is increased and new cell types are formed.

(4) The number of cells that make up the organism increases and the organism is capable of growth and repair.

40 ______

As the embryo develops, the cells of the ICM rearrange and change location. These changes are the result of Process *Z* and will give rise to all of the cell types needed to form structures such as nerves, organs, and muscles, such as the cardiac muscles of the heart. When a person suffers a heart attack, some of their cardiac muscle cells die. This causes permanent damage to the heart because adults cannot produce new cardiac cells under normal conditions. Doctors used animal models (pigs) to test several therapies aimed at repairing damaged heart tissue.

41. Describe a trade-off researchers must consider when using pigs as model organisms to cure heart disease in humans. [1]

___

___

___

Base your answers to questions 42 through 45 on the information below and on your knowledge of biology.

Driving the Florida Panther to Extinction

The decline in the Florida panther population is an example of how humans have had an impact on the biodiversity of an area. The panther is a large cat that preys primarily upon wild hogs, raccoons, and deer. They can be found in forested areas, pinelands, and freshwater swamp forests.

During the 1800s and early 1900s, habitat loss and hunting led to the panther's near-extinction. By the mid-1980s, only 20 to 30 panthers could be found in the wild. Conservation efforts to increase the panther population started with the release of eight females from Texas in 1995 into available panther habitat in south Florida, much of which was protected from human activity. The Florida panther population historically bordered the Texas population, with interbreeding occurring naturally. Such conservation efforts have succeeded in bringing the wild panther population up to about 200 individuals.

Florida Panther Territory

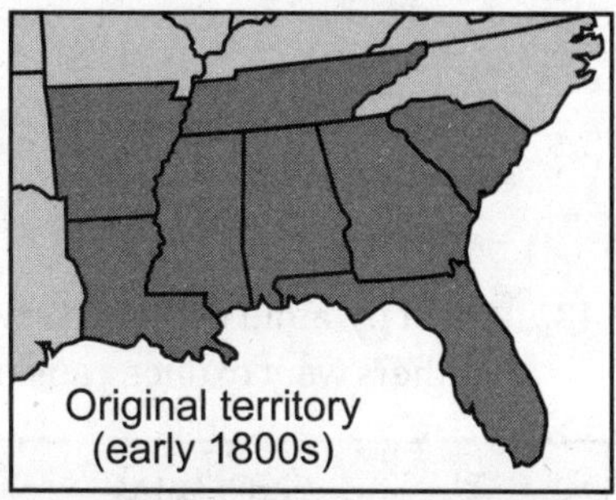

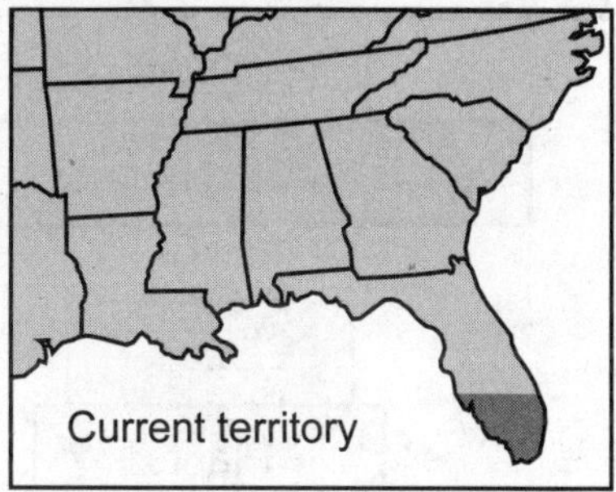

42. Which explanation regarding the Florida panther would best be supported by the evidence in the information provided?

(1) The average panther reproductive rate will increase in order to maintain the panther population.

(2) If their territory decreases because of housing developments, it will become easier for the males to find genetically diverse mates.

(3) Due to the size of the population, genetic variation is most likely low, decreasing the ability of the panther population to adapt to environmental changes.

(4) The panther population has rebounded and will most likely continue to rise, regardless of changes to their habitat.

42 ______

The model below represents a balanced Florida ecosystem.

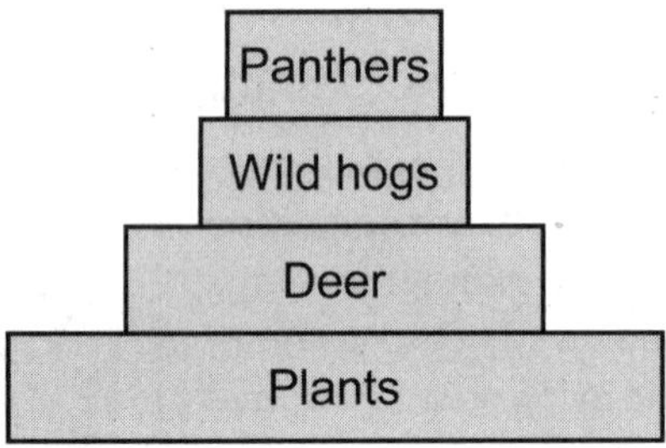

43. Which pyramid of biomass would best represent the Florida ecosystem if the panthers went extinct, causing a sudden imbalance in the ecosystem?

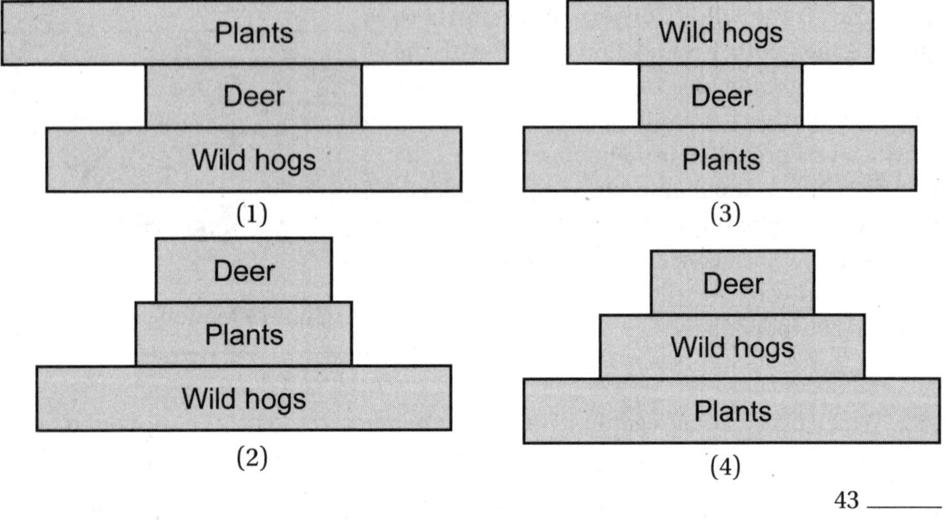

43 ______

Human-caused habitat fragmentation from the late 1800s through today has broken up the once-large panther territory. Urbanization has contributed to the fragmentation of the panther habitat in Florida.

The model shows some information about habitat fragmentation.

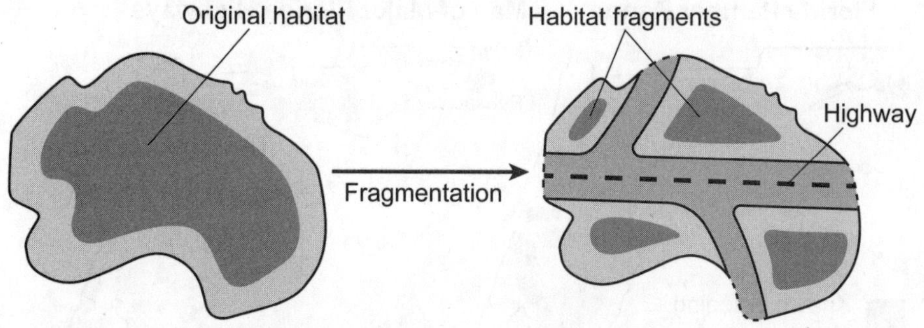

44. Which statement supports the claim that habitat fragmentation will increase the likelihood of the Florida panther becoming extinct?

(1) Habitat fragmentation can interfere with panther migration patterns and access to resources.

(2) Fragmentation will increase biodiversity that could lead to decreased resource availability.

(3) Competition between panthers living in different habitat fragments will increase.

(4) Panthers will be more likely to migrate to new habitats that lack feral hogs and raccoons.

44 ______

Some conservationists are now concerned that their original efforts to restore the Florida panther population may have been undone. They claim that new proposals to increase the number of housing developments and highways within the Florida panther breeding range will reduce the panther population to levels that occurred in the 1980s.

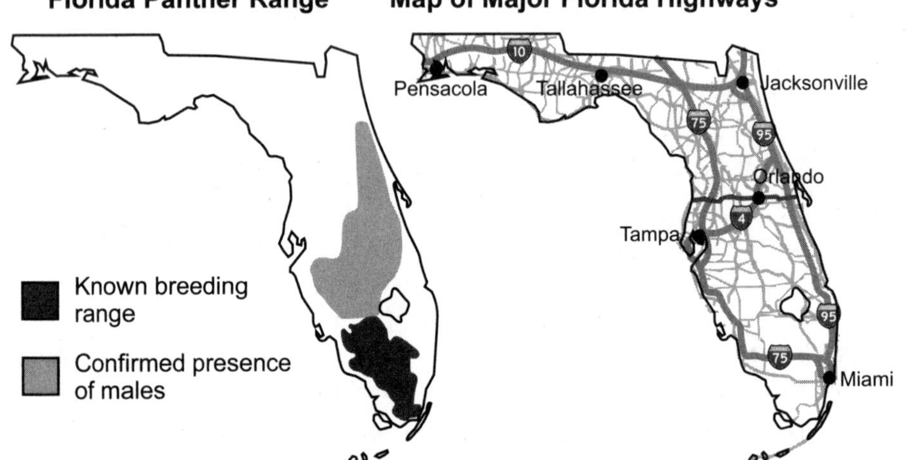

A proposed solution to reduce panther deaths caused by cars is to build wildlife bridges. When designing wildlife bridges, engineers in Florida prioritized the criterion of panther survival while also considering constraints such as cost, driver safety, environmental impacts, and aesthetics.

A Wildlife Bridge

45. Evaluate wildlife bridges as a solution to reduce the impact on panther populations, based on prioritized criteria *and* constraints. [1]

____

____

____

____

____

Answer Explanations August 2025

1. **Sample Response:** The vascular bundle connects the two systems: xylem moves water and minerals absorbed by the root system up to the shoot system (e.g., leaves) for photosynthesis, and phloem carries sugars made in the leaves through the shoot system down to the root system for use or storage.

   Explanation:

   This describes an interaction with both systems: roots supply water/minerals; shoots make food; and the vascular bundle transports in both directions so cells throughout the plant maintain homeostasis.

2. **(3)** Sap (which distributes sugars and water through phloem) together with root hairs (which absorb water and exchange gases) ensures that plant cells receive the materials needed for cellular respiration. Sugars delivered by sap are the fuel, and oxygen diffuses near roots and into tissues; together these inputs allow mitochondria to make ATP and maintain homeostasis.

   ## Wrong Choices Explained

   (1) Cellular respiration occurs in nearly all living plant cells (not only in root cells), and sap distributes materials to many tissues, not just roots.

   (2) Sugars are not made by combining sugar with carbon dioxide; photosynthesis makes sugar from carbon dioxide and water using light energy.

   (4) Root hairs and sap do not by themselves provide the full set of raw materials for photosynthesis to root cells. (Photosynthesis requires light and chloroplasts; most roots lack chlorophyll.)

3. **(1)** The production map shows optimal syrup production areas shifting northward under warming scenarios. Cooler conditions needed for sap sweetness and flow will occur farther north in the future; therefore, optimal production areas move north as climate warms.

   ## Wrong Choices Explained

   (2) The maps explicitly show a shift in optimal zones; they do not remain the same as 1950 to 1999.

   (3) Future optimal production is not predicted mainly at 38 to 42 degrees latitude; projections show movement poleward, not equatorward.

   (4) The figure does not support total loss of maple trees in northern areas or complete cessation of syrup production; it shows relocation of optimal zones.

4. **Sample Response:** Older and larger-diameter trees (including maples) store more carbon in their tissues and can sequester carbon at higher rates. As diameter increases (e.g., from 40 to 60 cm), total carbon stored rises substantially, moving carbon from the atmosphere (CO_2) into the biosphere (wood, roots, leaves).

   Explanation:

   Tree age/size influences both rate and total storage: maturing trees accumulate more biomass and thus more carbon, helping regulate atmospheric CO_2.

5. **(3)** Broadcast spawning releases millions of gametes, dramatically increasing the chance that some offspring survive to adulthood despite high mortality in nutrient-poor waters and cannibalism. Producing vast numbers is a reproductive strategy that offsets heavy early losses.

Wrong Choices Explained

(1) Disease spread is a cost of crowding; it does not explain a survival advantage of spawning.

(2) Food depletion is also a cost, not an advantage; the question asks how the behavior aids survival.

(4) Accumulation of wastes and low oxygen are disadvantages of crowding, not a reason the strategy improves survival.

6. **Sample Response:** Shrews invest in parental care—offspring nurse for approximately 3 to 4 weeks and the caravan behavior keeps young physically connected and protected when the nest is moved. This higher level of care and protection raises the probability that each offspring survives, so shrews can maintain their population even though each reproductive event produces far fewer offspring than tuna.

   Explanation:

   The trade-off is "quality over quantity": high parental investment raises survival rate per offspring, balancing low fecundity.

7. **(2)** Curve *C* represents very high early mortality with only a tiny fraction surviving to adulthood—consistent with tuna, where only approximately 2 of 30 million fertilized eggs reach adulthood. The key factor is that very few young survive to adulthood.

Wrong Choices Explained

(1) Releasing many gametes explains fecundity, but this alone doesn't define survivorship; the defining feature is the low number of offspring that survive.

(3) Migration at a certain size does not determine survivorship patterns across the lifespan.

(4) Cannibalism does not lead to approximately 50% survival; tuna survival to adulthood is far below 50%.

8. **Sample Response:** Curve *B* best represents shrews: about half of the offspring survive to adulthood due to parental care (nursing and caravan protection). Curve *A* suggests high survival to old age; Curve *C* suggests very high early mortality. Shrews fall in between—moderate juvenile mortality with significant adult survival.

Explanation:

Caravan behavior and nursing improve survival but do not eliminate early losses entirely.

9. **Sample Response:** Adult bison that form double protective rings around calves increase calf survival. Calves that survive to reproduce are more likely to pass on genes and learned behaviors associated with this protective strategy. Over generations, natural selection favors herds exhibiting this behavior, making it common in the population.

Explanation:

Behaviors that enhance survival and reproduction become more prevalent through differential reproductive success.

10. **(4)** A severe, unregulated population crash in the 1800s created a genetic bottleneck—few survivors contributed DNA to all modern bison. Modern populations therefore have reduced genetic diversity compared with ancestral herds prior to the 1800s.

Wrong Choices Explained

(1) More mates after the crash does not increase diversity lost during the bottleneck; variation must exist to be maintained.

(2) Random mutations occur but are too few to quickly restore ancestral diversity; the dominant effect of a bottleneck is loss of variation.

(3) Modern DNA is not identical to ancestral DNA; genetic drift and subsequent selection occurred, and diversity was lost.

11. **(1)** Cellular respiration in silkworm cells breaks the bonds in glucose obtained from mulberry leaves and converts that chemical energy into ATP, which is the usable energy currency for cellular processes throughout the silkworm's tissues.

Wrong Choices Explained

(2) Respiration occurs in the silkworm, not in the plant to provide amino acids as "cellular energy."

(3) Digestion does not break glucose into amino acids, and ATP is not "built" from amino acids.

(4) Digestion in mulberry leaves is not the relevant process; plants make glucose by photosynthesis, and silkworms use that glucose via respiration.

12. **Sample Response:** After the silkworm ingests mulberry macromolecules, its digestive system hydrolyzes them to subunits. The elements from glucose (carbon, hydrogen, oxygen) are recombined with nitrogen to form amino acids, which cells then link to build silkworm-specific proteins needed for growth and function.

Explanation:

This uses the idea of rearranging elements: glucose provides C, H, O; adding N yields amino acids; ribosomes synthesize proteins from amino acids.

13. **(2)** Receptors in the antennae (nervous system) detect mulberry scent molecules and send signals enabling the silkworm moth to locate leaves; the digestive system then breaks down leaf macromolecules into usable subunits. This coordination between sensory/nervous and digestive systems enables the silkworm to find and use the nutrient source.

Wrong Choices Explained

(1) Scent detection does not trigger cocoon spinning; cocooning is a distinct behavior and life-stage process.

(3) The nervous system does not command breakdown of fats specifically to "produce new muscle"; digestion breaks diverse macromolecules into subunits.

(4) Muscles do not "send messages" to the nervous system to receive scent; signals travel from receptors to the brain and onward to effectors.

14. **Sample Response:**

mRNA: CUA GUU AAU UUA
Amino acids: LEU – VAL – ASN – LEU

Explanation:

The DNA template GAT CAA TTA AAT transcribes to mRNA CUA GUU AAU UUA (complementary base pairing); codons translate to the listed amino acids using the universal codon chart.

15. **(2)** Between 1977 and 1989, the graph of ivory sales rises while the African elephant population graph declines—opposite trends consistent with poaching pressure from ivory demand. This supports the claim that elephants were being poached for their tusks.

Wrong Choices Explained

(1) The two graphs do not show a "similar trend" across 1977 to 2016; early periods show inverse trends.

(3) There is no evidence that predator pressure stopped; population changes align with human poaching pressures.

(4) Illegal poaching from 1979 to 1990 is not the key evidence stated; the most important evidence is the data relationship between ivory sales and population.

16. **(2)** From 1969 (10.5%) to 1989 (38.2%), the frequency of tusklessness in females increased—consistent with strong selection against tusks (poaching). This describes a clear increasing trend across those years.

Wrong Choices Explained

(1) 1988 to 1991 shows 32.6% → 33.3% (a slight increase, not a decrease).

(3) The total number of tuskless females did not remain constant from 1969 to 1993; counts and percentages changed.

(4) From 1989 to 1991, the total number of tuskless females actually decreased (63 → 57).

17. **(3)** Selective poaching of tusked elephants removes many tusked individuals from the gene pool, increasing the proportion of tuskless females (a heritable trait). Thus human activity (poaching for ivory) has increased the tuskless trait among females.

Wrong Choices Explained

(1) Global ivory sales decreasing would not specifically reduce tuskless males; tusklessness in males is lethal and not the driver here.

(2) Tighter enforcement does not increase tuskless male production; the allele is lethal to males.

(4) Demand for ivory would decrease, not increase, the frequency of tusked individuals; it selects against tusks, leaving more tuskless females.

18. **(1)** With more tuskless elephants, fewer trees are uprooted and bark stripped, allowing woody vegetation to increase. That shifts resource availability and habitat structure, potentially reducing grass available to other herbivores that rely on open areas—destabilizing the ecosystem balance.

Wrong Choices Explained

(2) Tuskless elephants dig fewer deep holes; they mainly use existing holes, so deep-hole availability would not increase.

(3) Less tree knocking and bark stripping would generally increase woody vegetation, not decrease it.

(4) Reduced uprooting would likely expose fewer soil organisms; vulnerability to predators would not necessarily increase.

19. **Sample Response:** Beehive fences are most effective with the least ecosystem impact. They deter elephants using a natural fear of bees, protecting crops and people without removing elephants from the ecosystem or reducing biodiversity. Managed hunting can preserve habitat and regulate numbers but directly removes animals and can reduce local biodiversity.

Explanation:

Evaluation weighs criteria: effectiveness at reducing conflict and minimal ecological impact. Beehive fences meet both; hunting adds risks (removal, enforcement burdens).

20. **(1)** Phytoplankton photosynthesize: they take in carbon dioxide and water from the hydrosphere/atmosphere and produce glucose and oxygen, moving carbon from the atmosphere/hydrosphere into the biosphere. Option (1) best represents this sphere-to-sphere transfer.

Wrong Choices Explained

(2) It reverses the direction of movement, showing glucose and oxygen going into phytoplankton and carbon dioxide and water coming out, which confuses photosynthesis with respiration.

(3) It suggests the wrong flux directions in relation to phytoplankton's role as producers.

(4) It does not correctly show carbon moving into biomass through photosynthesis.

21. **Sample Response:** Process *X* is decomposition. When organisms die, decomposers break down biomass and release carbon dioxide to the atmosphere (and dissolved CO_2 to water). This transfers carbon from the biosphere back to the atmosphere/hydrosphere, continuing the carbon cycle.

Explanation:

Decomposition links spheres: dead organisms (biosphere) → CO_2 in air/water (atmosphere/hydrosphere).

22. **Sample Response:** A rapid increase in krill would consume more phytoplankton, reducing photosynthesis and the uptake of atmospheric CO_2. With fewer phytoplankton drawing down CO_2, atmospheric carbon would rise. Although more krill fecal pellets could sink carbon, the immediate effect of reduced phytoplankton is less carbon fixation and less short-term removal from the atmosphere.

Explanation:

Short-term biosphere drawdown depends on phytoplankton abundance; heavy grazing suppresses primary production and CO_2 uptake.

23. **(3)** When krill eliminate wastes, dense carbon-rich fecal pellets sink to the seafloor and can remain buried for long periods. This transfers and stores carbon in ocean sediments (hydrosphere/geosphere), making elimination a carbon-sink process.

Wrong Choices Explained

(1) Photosynthesis removes, not releases, carbon to the atmosphere; and krill do not photosynthesize in air.

(2) Respiration releases CO_2 but does not store carbon; it is the opposite of a sink.

(4) "Acidification in the biosphere" is not a process that stores carbon; the main carbon sink is the sinking and burial of pellets.

24. **(3)** KRT1 and KRT12 have different DNA sequences and are expressed in different tissues. The sequence of DNA determines the amino acid sequence of each protein, which determines protein structure and function—explaining why keratin 1 (skin) and keratin 12 (cornea) differ.

Wrong Choices Explained

(1) mRNA structure does not determine the DNA code; DNA sequence determines mRNA and protein.

(2) Proteins do not contain the genetic information that determines DNA; information flows as DNA → RNA → protein.

(4) Protein structure does not determine DNA base order; causality is reversed.

25. **Sample Response:** All human somatic cells contain the same chromosomes and genes, including KRT1 (chromosome 12) and KRT12 (chromosome 17). Specialized cells result because different cells express different genes: skin keratinocytes express KRT1 to make keratin 1; corneal cells express KRT12 to make keratin 12.

Explanation:

Cell specialization arises from selective gene expression—turning specific genes "on" in certain tissues while others remain silent.

26. **Sample Response:** The mutated KRT10 sequence is missing two codons relative to the normal gene (e.g., Glu and Pro positions are deleted). This deletion shortens and alters the amino acid sequence, changing the protein's structure and impairing keratin production in skin cells.

Explanation:

Changes in nucleotide number/order change codons and thus amino acids, yielding an altered polypeptide with reduced or lost function.

27. **(2)** Genetically corrected keratinocytes divide by mitosis and pass the functional keratin gene to their daughter cells. Once a healthy population of skin cells is re-established, ongoing mitotic divisions maintain appropriate keratin production without modifying germ cells.

Wrong Choices Explained

(1) Keratinocytes do not undergo meiosis; meiosis occurs in gamete formation, not in skin maintenance.

(3) Unmodified skin cells still undergo mitosis; the point is that corrected cells can repopulate and restore function.

(4) Meiosis in body (somatic) cells does not occur; DNA in skin is not edited into the germline by this therapy.

28. **(2)** Keratin variation reflects inherited DNA sequence changes that were survivable and beneficial in each lineage; these genetic differences were passed on, leading to different keratin structures adapted for scales, claws, or fur.

Wrong Choices Explained

(1) Needs do not directly cause DNA to change; mutations arise randomly and are filtered by selection.

(3) They did not all share the same environment first and then mutate identically; divergence reflects different selection pressures.

(4) Keratin did not originate in crocodiles and then get "modified" in others; lineages inherit and diversify genes over time.

29. **Sample Response:** Blood from the circulatory system enters kidney capillaries, where nephrons of the excretory system filter wastes and reabsorb most water back to the blood. Long nephron loops create strong osmotic gradients that pull water from filtrate, producing very concentrated urine and conserving body water.

Explanation:

Interaction: capillaries (circulatory) deliver blood to nephrons (excretory); selective reabsorption returns water to the blood, minimizing water loss.

30. **(2)** Kangaroo rats have very long nephron loops and produce the highest urine:plasma ratio, showing strong water conservation—an adaptation favored by natural selection in deserts. Individuals with longer loops conserved more water and were more likely to survive and reproduce, spreading these traits.

Wrong Choices Explained

(1) Humans do not have the lowest ratio; beavers have the lowest, and nephron length correlates with the environment.

(3) There is a clear relationship: longer loops → higher maximal urine concentration.

(4) Nephron loop length varies with the environment; it is not the same across mammals.

31. **(4)** The nervous system detects the rattlesnake strike and rapidly signals the leg muscles; the muscular system contracts, launching the escape leap. This neural-muscular interaction enables the kangaroo rat's lightning-quick maneuver that often helps them evade the strike.

Wrong Choices Explained

(1) The respiratory system does not send neural signals to the spinal cord to trigger leaps.

(2) The circulatory system does not send neural signals to muscles; nerves do.

(3) Escape requires rapid contraction, not "slowing cellular respiration" prior to movement.

32. **Sample Response:** Extended, widespread drought lowers plant productivity and the moisture available in seeds that kangaroo rats eat. With less food and water in the habitat, the environment supports fewer individuals, so the carrying capacity for kangaroo rats decreases.

Explanation:

Carrying capacity (K) depends on resource availability; persistent drought reduces resources and thus lowers K.

33. **(3)** Fossils show ancestral whales with leg and pelvic bones; modern whales retain vestigial pelvic structures. This anatomical evidence supports descent from land-living mammals.

Wrong Choices Explained

(1) Early ancestors did not become more "adapted to land"—the trend is toward aquatic life.

(2) Counting species with non-functional pelvises does not by itself establish descent patterns.

(4) Learned behaviors are not inherited genetically; anatomical and genetic evidence is required.

34. **(4)** Whales (mammals) and sharks (fish) have similar streamlined body shapes because natural selection favored efficient swimming in both groups—an example of convergent evolution, not close genetic relatedness.

Wrong Choices Explained

(1) Aquatic organisms are not all closely related; similarity can arise independently under similar selection pressures.

(2) Marine mammals did not evolve directly from fish; they evolved from terrestrial mammals.

(3) Identical mutations are not required for similar forms; different genetic paths can yield similar phenotypes.

35. **Sample Response:** One correct tree (among several equivalent representations):

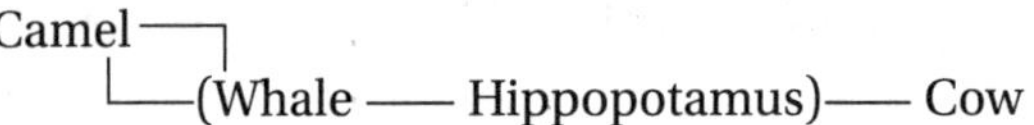

Examples of 1-credit responses:

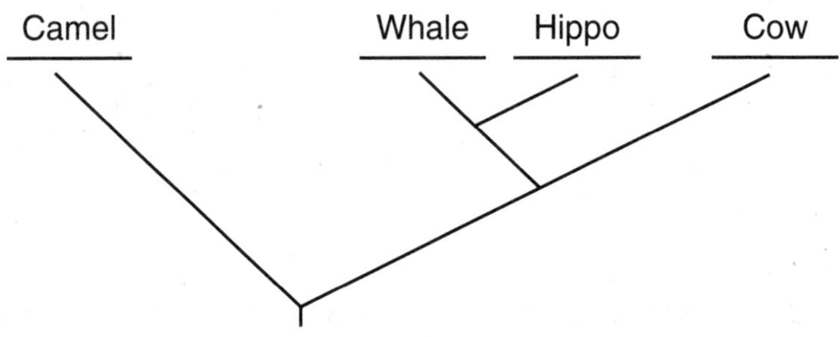

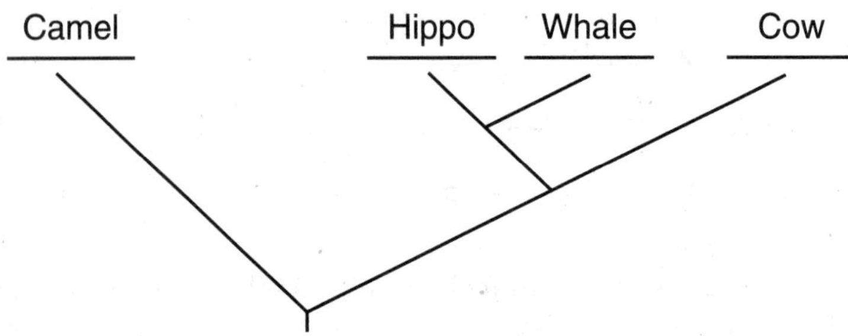

This places whale closer to hippo (shared transposon positions) than to camel and nests cow appropriately based on the transposon table.

Explanation:

Shared, uniquely placed transposons indicate common ancestry; taxa sharing more insertions at the same genomic locations are more closely related.

36. **(2)** DNA sequences and transposon locations provide high-resolution molecular evidence, and skeletal structures provide anatomical evidence. Combined, these data best reveal evolutionary relationships between goats and the listed mammals.

Wrong Choices Explained

(1) Habitat is ecological, not phylogenetic; embryonic development helps to determine evolutionary relationships, but DNA/transposons are stronger molecular evidence.

(3) Habitat and fossils help to determine evolutionary relationships, but without genetic data the resolution is limited.

(4) Coloration and simple bone counts lack the phylogenetic signal of DNA/transposons and detailed morphology.

37. **(1)** Process *X* (fertilization) produces a new single cell containing a full set of genetic information needed to form an embryo. Without combining haploid gametes, the zygote would lack the complete genome required for development.

Wrong Choices Explained

(2) Fertilization does not guarantee survival in any environment; it provides the starting genome.

(3) Process *X* does not ensure only favorable traits; it recombines parental genes, creating variation.

(4) Each parent contributes different genetic material; they do not pass on identical genomes.

38. **(2)** Process *Y* is mitosis. Mitosis produces daughter cells with the same chromosome number as the parent cell (46 → 46 → 46 . . .), maintaining the typical human diploid number throughout growth and development.

Wrong Choices Explained

(1) This shows meiosis-like reduction (23 ↔ 46) and does not represent mitotic cell division during development.

(3) This depicts reduction from 46 to 23, which is not mitosis.

(4) This sequence does not maintain consistent diploid numbers in all daughter cells.

39. Sample Response: Cell *A* contributes half of the DNA to the zygote. Because the morula forms by repeated mitosis of that zygote, every resulting cell inherits identical copies of the DNA—including the original error from Cell *A*.

Explanation:

Mitosis copies the genome faithfully (including any errors) to all daughter cells during early cleavage.

40. (4) Across all stages, Process *Y* (mitosis) increases the number of cells with the same genetic content, enabling growth and tissue repair. By expanding cell numbers while preserving the genome, organisms grow from a single cell to trillions.

Wrong Choices Explained

(1) Mitosis does not eliminate "damaged genes"; it copies existing DNA.

(2) Producing gametes is meiosis, not mitosis; the question refers to growth and development.

(3) Mitosis does not increase genetic variation among somatic cells; it preserves nuclear DNA identity.

41. Sample Response: Benefit: Pigs have organ size and physiology similar to humans, allowing realistic testing of cardiac therapies and reducing risk before human trials. Drawback (constraint): Pigs differ from humans in important ways (immune responses, metabolism), so results may not fully predict human outcomes; ethical and cost constraints also apply.

Explanation:

A valid trade-off pairs at least one benefit (translational relevance, safety data) with at least one drawback (species differences, ethics, cost).

42. (3) A small population that recently rebounded from near-extinction is likely to have low genetic variation. Low variation reduces the ability to adapt to environmental changes (disease, habitat shifts), so the population remains vulnerable despite protections.

Wrong Choices Explained

(1) Reproductive rate does not automatically increase to maintain population; it is constrained by biology and the environment.

(2) Smaller territories make it harder—not easier—to find genetically diverse mates.

(4) The population will not necessarily continue rising; habitat loss can reverse gains.

43. (3) If top predators (panthers) go extinct, their prey (deer, wild hogs) increase in biomass relative to plants, which can be overgrazed and decrease. The correct biomass pyramid would show reduced plant biomass supporting inflated herbivore biomass without the apex predator regulating them.

Wrong Choices Explained

(1) This arrangement would not reflect trophic imbalances caused by predator loss.

(2) Depicting plants above herbivores in biomass is inconsistent with primary producer dominance in mass and energy base.

(4) This pyramid misorders or misallocates trophic masses relative to predator removal outcomes.

44. (1) Fragmentation breaks continuous habitat into isolated patches separated by roads and development, disrupting migration routes and access to resources and mates. These barriers increase inbreeding and local extinctions, raising the overall risk of extinction.

Wrong Choices Explained

(2) Fragmentation usually decreases local biodiversity within patches and isolates populations; it does not increase biodiversity in a way that helps panthers.

(3) While competition can change, isolation mainly reduces movement, gene flow, and access to resources.

(4) Moving to entirely new habitats is unlikely; suitable connected habitat is limited, and barriers impede movement.

45. Sample Response: Wildlife bridges directly address the priority—panther survival—by reconnecting fragmented habitat and reducing vehicle collisions. Constraints include cost (construction/maintenance), driver safety during/after construction, environmental impacts of building in sensitive areas, and aesthetics. Despite these constraints, well-designed bridges increase safe crossings, maintain gene flow, and support long-term population stability.

Explanation:

A strong evaluation references the prioritized criterion (survival) and at least two constraints (e.g., cost, safety, environmental impact, aesthetics).

NOTES

NOTES

NOTES

NOTES

NOTES

NOTES

NOTES